Oceanic Internal Waves and Internal Tides in the East Asian Marginal Seas

Oceanic Internal Waves and Internal Tides in the East Asian Marginal Seas

Editors

SungHyun Nam
Xueen Chen

MDPI • Basel • Beijing • Wuhan • Barcelona • Belgrade • Manchester • Tokyo • Cluj • Tianjin

Editors

SungHyun Nam
Seoul National University
Korea

Xueen Chen
Ocean University of China
China

Editorial Office

MDPI
St. Alban-Anlage 66
4052 Basel, Switzerland

This is a reprint of articles from the Special Issue published online in the open access journal *Journal of Marine Science and Engineering* (ISSN 2077-1312) (available at: https://www.mdpi.com/journal/jmse/special_issues/internal_tides).

For citation purposes, cite each article independently as indicated on the article page online and as indicated below:

LastName, A.A.; LastName, B.B.; LastName, C.C. Article Title. *Journal Name* **Year**, *Volume Number*, Page Range.

ISBN 978-3-0365-4213-3 (Hbk)
ISBN 978-3-0365-4214-0 (PDF)

Contents

About the Editors

SungHyun Nam

SungHyun Nam (Associate Professor) received his B.S., M.S., and Ph.D. degrees in Oceanography (Physical Oceanography) in 1999, 2001, and 2006 respectively all from Seoul National University (SNU). After earning his Ph.D degree he worked at Agency for Defense Development (ADD) for the ROK Navy for 2 years from 2006 to 2008. Then he moved to San Diego, CA, USA to work at the Scripps Institution of Oceanography (SIO), University of California, San Diego (UCSD) as a postdoctoral researcher. Until he returned to his mother-university (SNU), he stayed in San Diego for 6 years conducting his sea-going studies as postdoctoral researcher and assistant scientist. In 2014, he joined SNU as an assistant professor. He is now an associate professor at the SNU, and serves as a member of ROK governmental and international committees.

Xueen Chen

Xueen Chen (Professor) received his B.S. and M.S. degrees in Oceanography (Physical Oceanography) in 1994 and 1997 respectively both from Ocean University of China (OUC), and Ph.D. degree in Oceanography in 2004 from Hamburg University, Germany. He worked at OUC from 1997 up to now, during which he visited Hamburg University, Germany from 2002 to 2006, as PhD researcher and postdoctoral researcher, successively. He is now a professor in physical oceanography at OUC, and serves as Chairman of IASPO committee of China.

Journal of
Marine Science and Engineering

MDPI

Editorial

Oceanic Internal Waves and Internal Tides in the East Asian Marginal Seas

Sunghyun Nam [1,*] and Xueen Chen [2]

1 College of Natural Sciences, Seoul National University, Seoul 08826, Korea
2 College of Oceanic and Atmospheric Sciences, Ocean University of China, Qingdao 266100, China; xchen@ouc.edu.cn
* Correspondence: namsh@snu.ac.kr

1. Introduction

Inertia-gravity waves or internal waves (IWs) are ubiquitous in the stratified, rotating ocean. These waves are present at all frequencies between the local inertial frequency and buoyancy frequency, sometimes including diurnal and semidiurnal tidal frequencies. The two types of low-frequency IWs that originate from atmospheric disturbances (such as sea surface winds) and ocean tides (e.g., interactions between barotropic tides and bottom topography) are near-inertial waves (NIWs) [1] and internal (or baroclinic) tides (ITs) [2]. The energy of low-frequency IWs at larger scales transfers to smaller scales at higher frequencies. Continuum-frequency waves provide an illustration of this principle, where the characteristics of nonlinear internal waves (NLIWs) or internal solitary waves (ISWs) show that they will ultimately break and dissipate. IWs are known to play a key role in the redistribution of heat, momentum, and materials via turbulent mixing, such as diapycnal mixing [3]. It is clear to researchers that IWs may affect local and global climates [4], biogeochemistry and biological productivity [5], marine engineering and submarine navigation [6], and underwater acoustics [7].

Despite decades of study on IWs in other regions, the scientific understanding of IWs in the East Asian marginal seas is lacking detail. There exists a need for greater clarity on the mechanisms that underlie wave generation, propagation, evolution, and dissipation. In this editorial, we introduce eight publications of this Special Issue [8–15]. The wave types and regional waters covered in this review include ISWs/NLIWs in the Bali Sea; ITs, NIWs, ISWs/NLIWs in the South China Sea; ISWs/NLIWs in the East China Sea; and NIWs in the East Sea (also known as the Japan Sea). We discuss and collate the combined findings, with the aim of improving the understanding of the physical mechanisms of IWs (specifically the characteristics that define wave generation, propagation, and/or acoustic impacts).

Citation: Nam, S.; Chen, X. Oceanic Internal Waves and Internal Tides in the East Asian Marginal Seas. *J. Mar. Sci. Eng.* **2022**, *10*, 573. https://doi.org/10.3390/jmse10050573

Received: 11 April 2022
Accepted: 21 April 2022
Published: 23 April 2022

2. Recent Findings in the East Asian Seas

The ISWs/NLIWs are a great threat to submarine navigation as discussed in [8]. In the early morning of 21 April 2021 (local time), the Indonesian Navy Submarine (KRI Nanggala-402) crashed to the seafloor. The authors of [8] analyzed the ISWs/NLIWs in and around the submarine wreck site in the Bali Sea. They surveyed satellite remote sensing data collected from 12 April to 21 April and found that the ISWs/NLIWs had travelled across the deep basin of the Bali Sea [8]. The pathway of the waves passed through the submarine wreck site and then shoaled onto the continental shelf (originating from the Lombok Strait).

Several papers have investigated the three types of IWs found in the South China Sea [9–13]. The study completed by the authors of [9] applied a general ocean circulation model to the ITs in the Luzon Strait. Their work showed the impact of fortnightly stratification variability (as induced by tide–topography interactions) on the generation of ITs. Their contribution has led to a better understanding of the energy transfer between barotropic and baroclinic tides and shown a lead–lag relationship between barotropic tidal forcing and maximum baroclinic response within the fortnightly tidal cycle.

Using hybrid coordinate ocean model re-analysis, the authors of [10] reproduced the NIWs induced by Typhoon Megi in 2010 in the South China Sea. Their results revealed that typhoon-induced NIWs could propagate to a depth of 1000 m. They found that the damping and modal content of typhoon-induced NIWs were site-dependent. The results of their modelling showed that the first three baroclinic modes dominated and damped quickly in the region near the typhoon track, while the e-folding time of typhoon-induced NIWs could be longer than 20 days and higher modes were enhanced several days after the typhoon passage.

The three studies published by the authors of [11–13] addressed high-frequency IWs, particularly ISWs/NLIWs at frequencies close to the buoyancy frequency in the South China Sea. The study published by the authors of [11] investigated the distribution and source sites of ISWs/NLIWs northeast of Hainan Island, using satellite remote sensing data and a wave-front propagation model. Their work identified two types of ISWs/NLIWs originating from the Luzon Strait, spaced at both semidiurnal (northern region) and diurnal (southern region) tidal periods. On 22 May 2011, northeast of Dong-Sha Atoll, the authors of [12] observed strong ISWs/NLIWs with unprecedently large velocities (a peak westward velocity of 2.94 m/s and a peak downward velocity of 0.63 m/s), as measured by shipboard velocity observations. They inferred the wave's amplitude (~97 m) from backscatter observations and propagation speed (1.76 m/s estimated theoretically and 1.59 m/s inferred from the satellite remote sensing data). In further work conducted in the South China Sea [13], the authors reported on the substantial influence of ISWs/NLIWs on underwater sound propagation and ambient noise. Their paper described a passive acoustic monitoring and warning method for the strong velocity induced by ISWs/NLIWs, given that the power spectra of noise generated by the waves at frequencies below 100 Hz was almost 20 dB higher than ambient noise.

The authors of [14] developed a method to estimate the propagation speed and direction of ISWs/NLIWs using shipboard underway and moored observations. Their work applied two methods to estimate propagation speed and direction: apparent observations from a moving ship using the Doppler shift method (measuring change in frequency relative to the distance of the waves from the ship), and the time lag method (observing the distance between two locations of the wave at different times). The authors developed an optimal approach that then was applied to two cases of ISWs/NLIWs as observed in the northern regions of the East China Sea in May 2015 and August 2018.

The work published by the authors of [15] utilized a 21-year-long dataset of moored observations in the southwestern region of the East Sea (the Japan Sea), and focused on the non-seasonal (intra-seasonal, interannual, and decadal) variability of NIW kinetic energy far below the surface mixed layer. Their results identified nine periods of relatively high (*period high*), and seven periods of relatively low (*period low*) NIW kinetic energy. The work statistically revealed that the NIW kinetic energies in specific years and decades were significantly higher than those in other years and decades, in association with mesoscale circulation—NIW kinetic energy was enhanced/favored under conditions of negative relative vorticity and strong total strain.

3. Conclusions

This Special Issue contributes to a better understanding of the types of IWs in the East Asian marginal seas. We paid particular attention to descriptions of the processes that underly the generation, propagation, and/or acoustic impacts of different IWs. However, further aspects of regional IWs remain poorly understood, including specific details on the generation, propagation, evolution, breaking, and dissipation of NIWs, ITs, continuum-frequency waves, and ISWs/NLIWs.

Continuing to advance the scientific knowledge of regional IWs will aid researchers and environmental managers in addressing many local and global concerns, including issues related to climate, biogeochemistry, biological productivity, marine engineering, submarine navigation, and underwater acoustics. We hope that this Special Issue will be of interest to researchers in these fields, and to those scientists who are continuing to work on the impacts of IWs in the East Asian Seas.

Author Contributions: Writing—original draft preparation, S.N.; writing—review and editing, S.N. and X.C. All authors have read and agreed to the published version of the manuscript.

Funding: This research was a part of the project titled 'Deep Water Circulation and Material Cycling in the East Sea (20160040)' and 'Construction of Ocean Research Station and their Application Studies', funded by the Ministry of Oceans and Fisheries, Republic of Korea. This work is supported by the National Research Foundation (NRF), funded by the Korean Ministry of Education (NRF-2019R1A2C2085461).

Institutional Review Board Statement: Not applicable.

Informed Consent Statement: Not applicable.

Data Availability Statement: Not applicable.

Conflicts of Interest: The authors declare no conflict of interest.

References

1. Alford, M.H.; MacKinnon, J.A.; Simmons, H.L.; Nash, J.D. Near-Inertial Internal Gravity Waves in the Ocean. *Annu. Rev. Mar. Sci.* **2016**, *8*, 95–123. [CrossRef] [PubMed]
2. Vic, C.; Naveira Garabato, A.C.N.; Green, J.A.M.; Waterhouse, A.F.; Zhao, Z.; Melet, A.; De Lavergne, C.; Buijsman, M.C.; Stephenson, G.R. Deep-Ocean Mixing Driven by Small-Scale Internal Tides. *Nat. Commun.* **2019**, *10*, 2099. [CrossRef] [PubMed]
3. Müller, P.; Briscoe, M. Diapycnal Mixing and Internal Waves. *Oceanography* **2000**, *13*, 98–103. [CrossRef]
4. MacKinnon, J.A.; Alford, M.H.; Ansong, J.K.; Arbic, B.K.; Barna, A.; Briegleb, B.P.; Bryan, F.O.; Buijsman, M.C.; Chassignet, E.P.; Danabasoglu, G.; et al. Climate Process Team on Internal Wave-Driven Ocean Mixing. *Bull. Am. Meteorol. Soc.* **2017**, *98*, 2429–2454. [CrossRef] [PubMed]
5. Hung, J.-J.; Wang, Y.-H.; Fu, K.-H.; Lee, I.; Tsai, S.-S.; Lee, C.-Y.; Lu, W.-T.; Shen, Y.-J.; Lin, Y.-H. Biogeochemical Responses to Internal-Wave Impacts in the Continental Margin off Dongsha Atoll in the Northern South China Sea. *Prog. Oceanogr.* **2021**, *199*. [CrossRef]
6. Huang, X.; Chen, Z.; Zhao, W.; Zhang, Z.; Zhou, C.; Yang, Q.; Tian, J. An Extreme Internal Solitary Wave Event Observed in the Northern South China Sea. *Sci. Rep.* **2016**, *6*, 30041.
7. Kuperman, W.A.; Lynch, J.F. Shallow-Water Acoustics. *Phys. Today* **2004**, *57*, 55–61. [CrossRef]
8. Wang, T.; Huang, X.; Zhao, W.; Zheng, S.; Yang, Y.; Tian, J. Internal Solitary Wave Activities near the Indonesian Submarine Wreck Site Inferred from Satellite Images. *J. Mar. Sci. Eng.* **2022**, *10*, 197. [CrossRef]
9. Zhang, Z.; Chen, X.; Pohlmann, T. The Impact of Fortnightly Stratification Variability on the Generation of Baroclinic Tides in the Luzon Strait. *J. Mar. Sci. Eng.* **2021**, *9*, 703. [CrossRef]
10. Cao, A.; Guo, Z.; Pan, Y.; Song, J.; He, H.; Li, P. Near-Inertial Waves Induced by Typhoon Megi (2010) in the South China Sea. *J. Mar. Sci. Eng.* **2021**, *9*, 440. [CrossRef]
11. Liang, J.; Li, X.-M.; Fan, K. Distribution and Source Sites of Nonlinear Internal Waves Northeast of Hainan Island. *J. Mar. Sci. Eng.* **2022**, *10*, 55. [CrossRef]
12. Xu, A.; Chen, X.; Strong, A. A Strong Internal Solitary Wave with Extreme Velocity Captured Northeast of Dong-Sha Atoll in the Northern South China Sea. *J. Mar. Sci. Eng.* **2021**, *9*, 1277. [CrossRef]
13. Li, J.; Shi, Y.; Yang, Y.; Huang, X. Noise of Internal Solitary Waves Measured by Mooring-Mounted Hydrophone Array in the South China Sea. *J. Mar. Sci. Eng.* **2022**, *10*, 222. [CrossRef]
14. Lee, S.-W.; Nam, S. Estimation of Propagation Speed and Direction of Nonlinear Internal Waves from Underway and Moored Measurements. *J. Mar. Sci. Eng.* **2021**, *9*, 1089. [CrossRef]
15. Noh, S.; Nam, S. Nonseasonal Variations in near-Inertial Kinetic Energy Observed Far below the Surface Mixed Layer in the Southwestern East Sea (Japan Sea). *J. Mar. Sci. Eng.* **2022**, *10*, 9. [CrossRef]

Journal of
Marine Science and Engineering

MDPI

Article

Internal Solitary Wave Activities near the Indonesian Submarine Wreck Site Inferred from Satellite Images

Tongxin Wang [1], Xiaodong Huang [1,2,3,*], Wei Zhao [1,2,3], Shihao Zheng [4], Yunchao Yang [1] and Jiwei Tian [1,2,3]

1 Frontier Science Center for Deep Ocean Multispheres and Earth System (FDOMES) and Physical Oceanography Laboratory, Ocean University of China, Qingdao 266100, China; wangtongxin@stu.ouc.edu.cn (T.W.); weizhao@ouc.edu.cn (W.Z.); yunchaoyang@163.com (Y.Y.); tianjw@ouc.edu.cn (J.T.)
2 Sanya Oceanographic Institution, Ocean University of China, Sanya 572000, China
3 Qingdao National Laboratory for Marine Science and Technology, Qingdao 266100, China
4 Troop 92682, People's Liberation Army, Zhanjiang 524000, China; shihao_zheng@126.com
* Correspondence: xhuang@ouc.edu.cn; Tel.: +86-053266783912

Abstract: In the early morning of 21 April 2021 local time, the Indonesian Navy submarine *KRI nanggala-402* crashed in the Bali Sea (BS). As internal solitary waves (ISWs) are a great threat to submarine navigation, this paper analyzes the characteristics of ISWs in the BS by surveying satellite remote sensing images collected from 12–21 April 2021. The satellite images revealed active ISWs in the BS near the submarine wreck site with crest lengths approaching 200 km. Originating from the Lombok Strait (LS), the waves travelled northwestward across the BS deep basin, passed through the submarine wreck site, and shoaled onto the continental shelf west of the Kangean Islands, during which process, the propagation speed reached 2.69 m/s in the deep basin and 0.71 m/s in the shallow water. Based on the satellite images, the wave amplitude near the wreck site was reconstructed to be 41 m, and the reconstructed underwater wave structure showed a maximum vertical velocity of 10 cm/s. Satellite images also demonstrated the near-source evidence of ISWs near the Nusa Penida sill of the LS, and their generation were estimated to be related to the southward tidal current troughs.

Keywords: *KRI nanggala-402* submarine wreck; Lombok Strait; Bali Sea; internal solitary waves; remote sensing images

Citation: Wang, T.; Huang, X.; Zhao, W.; Zheng, S.; Yang, Y.; Tian, J. Internal Solitary Wave Activities near the Indonesian Submarine Wreck Site Inferred from Satellite Images. *J. Mar. Sci. Eng.* **2022**, *10*, 197. https://doi.org/10.3390/jmse10020197

Academic Editor: Lev Shemer

Received: 7 December 2021
Accepted: 26 January 2022
Published: 1 February 2022

1. Introduction

Internal solitary waves (ISWs) are the nonlinear short-period waves widely distributed in the global oceans [1–8]. The largest observed ISW has an amplitude of 240 m and its vertical current reaches 0.64 m/s [9]. Featuring large amplitudes and strong currents, ISWs in the oceans are considered as one of the major threats to submarine navigation [10]. Specifically, ISWs have the ability to cause strong density disturbances that can suddenly decrease the buoyancy of a submarine, resulting in a large depth drop in a very short time. On the other hand, the strong downward current in front of large amplitude ISW can exert a huge force on the submarine and may drag it to seabed [11,12]. The well-known USS Thresher nuclear submarine disaster in 1963 was possibly caused by internal waves [13].

Connecting the Bali Sea (BS) with the Indian Ocean, the Lombok Strait (LS) features steep bottom topographies, and its southern portion is occupied by the Nusa Penida sill with an average depth of 200 m. Research based on Synthetic Aperture Radar (SAR) images [14–17] and in-situ observation [18] have revealed active ISWs around the LS area, which were generated by tidal currents flowing over the Nusa Penida sill. Recently, comprehensive high-frequency observations in the LS have captured almost continuously internal wave packets with a maximum amplitude of approximately 40 m [19]. The northward-propagating ISWs would propagate into the BS and travel across the entire basin with an average speed of about 2.0 m/s, and their crests could extend for hundreds of kilometers [17]. Numerical simulation results also showed that the occurrences of ISWs in the LS area varied significantly over monthly and interannual timescales under the modulation of thermocline structure adjustment, monsoons and the Indonesian Throughflow [20–22].

In the early morning of 21 April 2021 local time (near 4:30 AM), the Indonesian Navy's submarine (*KRI nanggala-402*) crashed in the BS and all 53 crew members were died. Public information reported that the submarine *KRI nanggala-402* crashed about 60 miles north of Bali Island, at a water depth of ~850 m. So far, Indonesian officials have not announced the specific cause of the submarine wreck. On grounds of the abundant ISW activities in the BS, it is necessary to investigate the ISW characteristics around the time when the *KRI nanggala-402* was wrecked, which will be helpful to clarify the reasons of the submarine incident. Fortunately, satellites photographed dense ISW signals over BS from 12–21 April 2021. In this study, we collected the optical remote sensing images covering BS during those days and investigated the distribution, propagation and underwater structure of ISWs around the time of the *KRI nanggala-402* wreck.

2. Data and Methods

2.1. Satellite Images

On account of the changes of sea surface roughness induced by convergence and divergence in the wave front and rear, ISWs are often manifested as bright and dark stripes in optical remote sensing images [23]. Accordingly, remote sensing images are regarded as a useful tool to investigate ISWs in the oceans. Zhao et al. analyzed the polarity transition of ISWs over the continental shelf of the northern South China Sea using SPOT-3 satellite optical image [24], and Jackson utilized Moderate Resolution Imaging Spectroradiometer (MODIS) images for global internal wave detection [25]. Furthermore, based on MODIS images, Huang and Zhao extracted the characteristic parameters of a typical ISW in the deep water of northern South China Sea [26] and Ning et al. further established a model that is able to derive the amplitude of ISWs [27].

The optical remote sensing images employed in this paper were acquired by the MODIS sensor equipped on the National Aeronautics and Space Administration's (NASA's) Terra/Aqua satellite and the Visible infrared Imaging Radiometer (VIIRS) sensor equipped on the NOAA/Suomi NPP satellite. The MODIS data are obtained in 36 visible and infrared bands with a spatial resolution between 250 m and 1 km, which depend on acquisition wavelength, and the VIIRS data are obtained from 22 channels at two resolutions, 375 m and 750 m.

A total of 10 satellite images with distinguishable ISWs over the BS were collected (Figures 1 and 2) and they were taken from two durations, either from 9:30 to 10:00 or from 12:00 to 13:00 local time. The 10 images involve six days of the period from 12 to 21 April, and there were two images on 12, 14 and 19 April photographed with an interval of 175 min. Unfortunately, no satellite images were available on 20 April.

2.2. ISW Underwater Structure Reconstruction

ISW amplitude η_0 can be derived from satellite images using the following equation [28]:

$$\eta_0 = \frac{12\beta}{\alpha l^2} = 1.32^2 \cdot \frac{12\beta}{\alpha D^2}, \tag{1}$$

where D is the distance between the center of light and that of dark stripes, and l is the half-wavelength of the ISW. α and β are nonlinear coefficient and dispersion coefficient, in the Korteweg–de Vries (KdV) equation, and they are calculated from $\alpha = 3\int_{-H}^{0} c_0 \left[\frac{\partial\phi(z)}{\partial z}\right]^3 dz / 2\int_{-H}^{0} \left[\frac{\partial\phi(z)}{\partial z}\right]^2 dz$ and $\beta = c_0 \int_{-H}^{0} [\phi(z)]^2 dz / 2\int_{-H}^{0} \left[\frac{\partial\phi(z)}{\partial z}\right]^2 dz$, respectively. Here, $\phi(z)$ is the first-mode vertical eigenfunction of the wave and governed by Sturm–Liouville equation:

$$\frac{d^2\phi}{dz^2} + \frac{N^2(z)}{{C_0}^2}\phi = 0, \tag{2}$$

with the boundary conditions $\phi(0) = \phi(-H) = 0$, where C_0 is the eigenspeed of the Equation (2) and $N^2(z)$ represents the background stratification obtained from monthly climatological density profiles from the WOA18 (World Ocean Atlas 2018) dataset. This method has been proved to be reliable by comparing the inversion results of ISW amplitude from MODIS images in the northern South China Sea with the wave amplitude measurements from mooring observations [26].

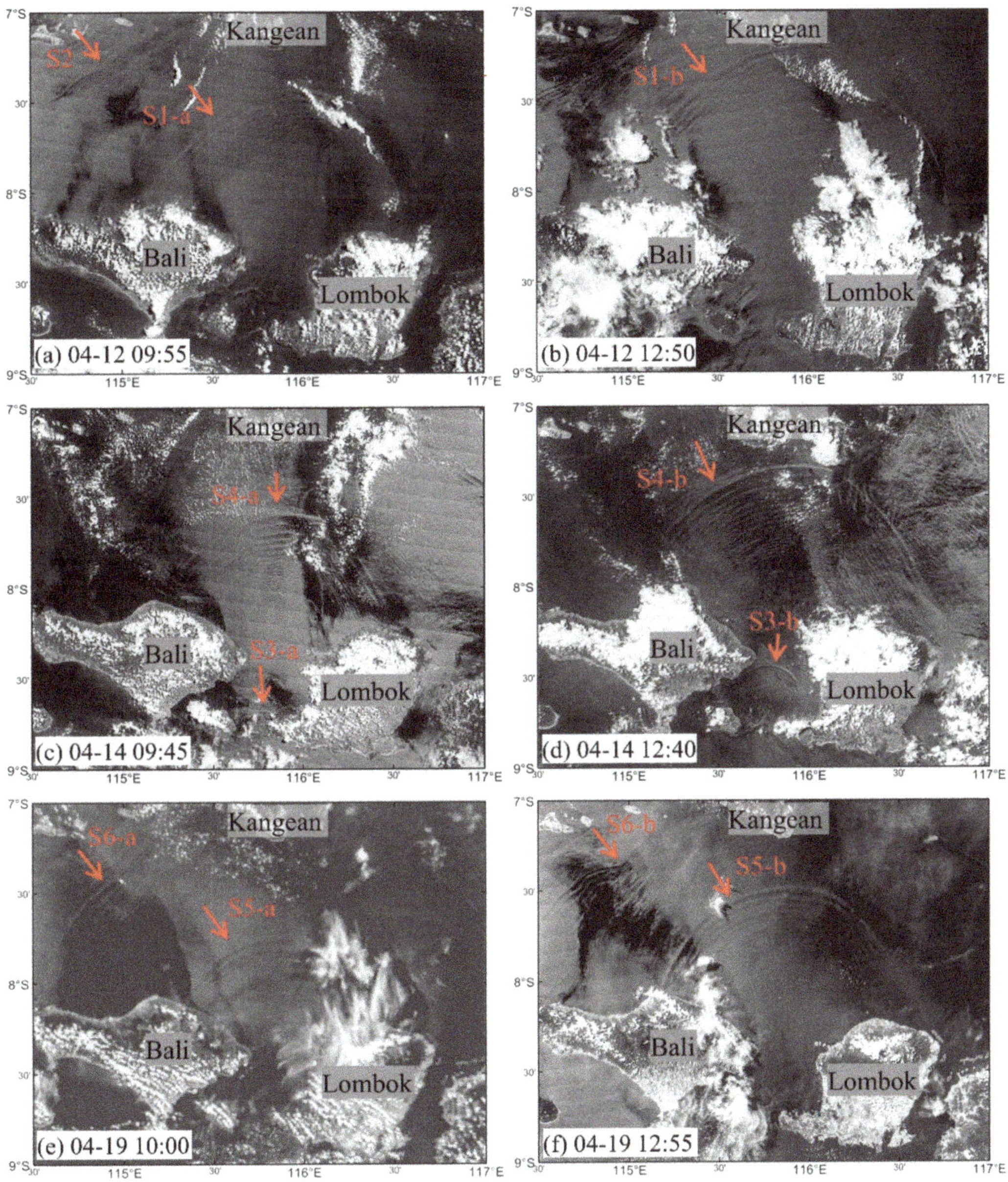

Figure 1. Optical remote sensing images over the BS acquired on 12, 14, and 19 April 2021 (local time) by MODIS.

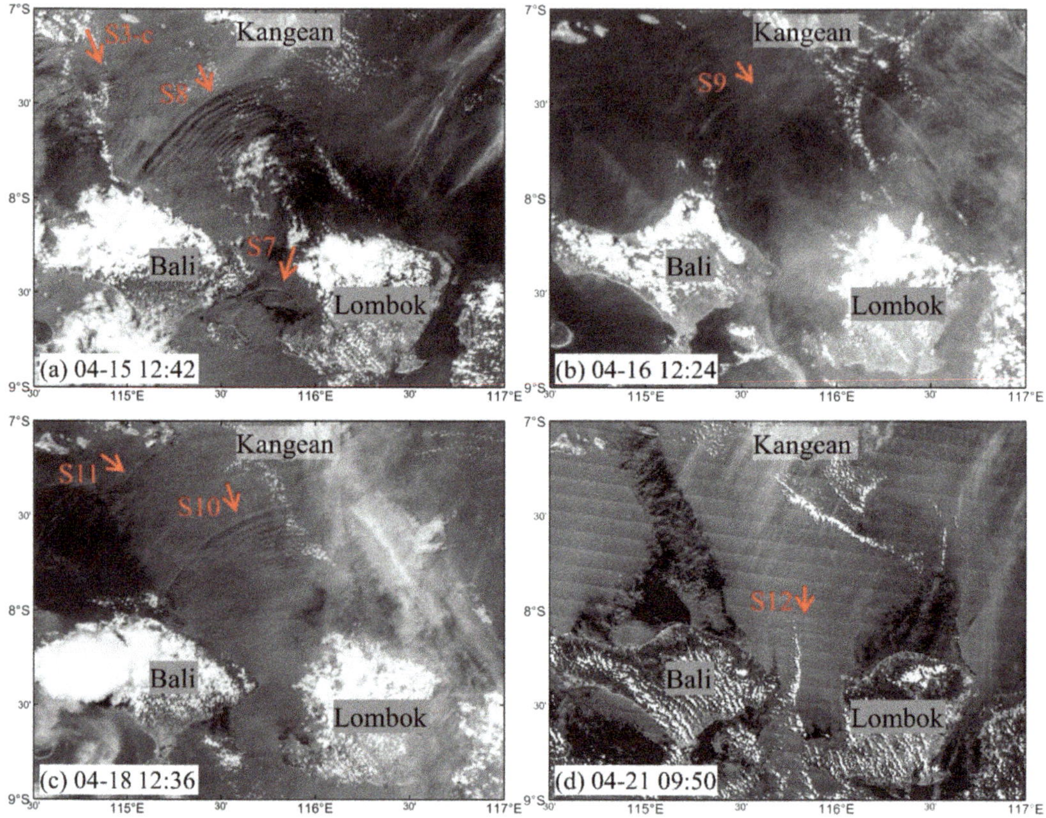

Figure 2. Optical remote sensing images over the BS acquired on 15, 16, 18 and 21 April 2021 (local time) by VIIRS (**a–c**) and by MODIS (**d**).

The waveshape of the ISW with amplitude η_0 can be obtained based on the solution to the KdV equation:

$$\eta(x,z) = \eta_0 sech^2 x \cdot \phi(z). \tag{3}$$

As an ISW arrives, the isopycnal surface, initially at depth z, is depressed to the depth $z' = z + \eta$. After rotating the horizontal axis of the coordinates to the ISW propagation direction, the continuity equation is written as:

$$\frac{\partial u(x,z')}{\partial x} + \frac{\partial w(x,z')}{\partial z'} = 0, \tag{4}$$

where u and w are the horizontal current along the wave propagation direction and the vertical current, respectively. The vertical velocity is regarded as the partial derivative of isopycnal displacement with respect to time, and thus $w(x,z') = \partial\eta(x,z)/\partial t$ and the first derivative of isopycnal depth in the vertical direction is given as $\partial z'/\partial z = 1 + \partial\eta/\partial z$. Therefore, local along-isopycnal horizontal current can be calculated from:

$$u(x,z') = -c\frac{\partial\eta(x,z)}{\partial z'} = -\frac{c\frac{\partial\eta(x,z)}{\partial z}}{1+\frac{\partial\eta(x,z)}{\partial z}}. \tag{5}$$

3. Results and Discussion

3.1. Spatial Distribution of ISWs

Figures 1 and 2 show dozens of ISWs, whose crests appear as northward convex arcs, covering the vast region of the BS from the southern LS, to the shallow continental shelf, to the west of the Kangean Islands. The crests of those ISWs almost bordered Bali Island on the left and Lombok Island on the right in the LS, and they diverged significantly during the northward propagation in the BS, passing through almost the entire basin. For the wave (S4-b) observed from the MODIS image taken at 12:40 on 14 April (Figure 1d), its crest line spanned about 1.5 degrees of longitude in the BS basin with a length of nearly 200 km.

Moreover, most ISWs in the BS appeared in the form of multi-wave packets that contain a number of rank-ordered solitons. For example, there existed more than 20 solitons in the packet S4-b, and those solitons spanned more than 60 km along their propagation direction and filled nearly half of the area between the LS and the Kangean Islands. The distances between the solitons in the ISW packets decreased from the front to the rear of the packet, and solitons were highly concentrated in the packet rear. Comparisons between Figure 1c,d also demonstrate the evolution process of S4, featuring the newborn solitons in the packet rear during propagation.

Up to three multi-wave ISW packets appeared simultaneously in one satellite image. In the VIIRS image of 15 April (Figure 2a), three distinguishable wave packets (S7, 8 and 3-c) were distributed in the LS, the BS basin and the continental shelf to the west of Kangean Islands, respectively, occupying a large portion of the BS. Actually, in the majority of the satellite images presented in Figures 1 and 2, two multi-wave ISW packets are easily seen, suggesting the prevailing solitary waves in the BS near the time of the submarine wreck.

The locations of all leading ISW crests in 10 remote sensing images were extracted and plotted together to exhibit the distribution of ISWs (coloured curves in Figure 3a). It can be seen that, near the time of the submarine wreck, ISWs generally propagated toward the northwest, and thus, the ISWs were concentrated in the area north of Bali Island where the submarine wreck occurred. However, in the SAR images collected by Karang et al. from 2006 to 2011 and Karang et al. from 2014 to 2015 [14,29], most of ISWs propagated northeastward after being emitted from the LS, and those waves were mainly distributed in the area to the north of the Lombok Island rather than the Bali Island. This phenomenon indicates that the spatial distribution of ISWs in the BS has significant temporal variability, which may be modulated by dynamic processes such as mesoscale eddies. As is clear to all, ISWs in the area to the north of Bali Island were extraordinarily active in April 2021, which significantly increased the likelihood that submarines would encounter internal waves.

3.2. Propagation Speed of ISWs

Two MODIS images were available on 12, 14 and 19 April (Figure 1), and the 175-min interval of imaging time makes it possible to accurately calculate the propagation speed of ISWs (Figure 3c).

In the LS, a clear ISW packet (S3) was observed from the MODIS images on 14 April (Figure 1c,d). Within 175 min, the leading part of the wave center propagated northward for ~27 km, and the average speed was calculated to be 2.54 m/s accordingly. This value is very close to the ISW speed of 2.5 m/s obtained by Lindsey et al. [30] with a shorter 10-min time steps. Moreover, based on the assumption that the time interval of ISW generation in the LS is consistent with the tidal cycle, Susanto et al. [17] estimated the speeds of northward-propagating ISWs from the Lombok Strait as 1.97 and 1.96 m/s using ERS-1/2 SAR images of 23 and 24 April 1996 and Karang et al. [29] obtained a speed of 2.05 m/s using Landsat 8 image of 17 May 2015. It can be seen that the speed of ISW in the LS is not the same at different periods, and here we show a variation range of about 0.5 m/s.

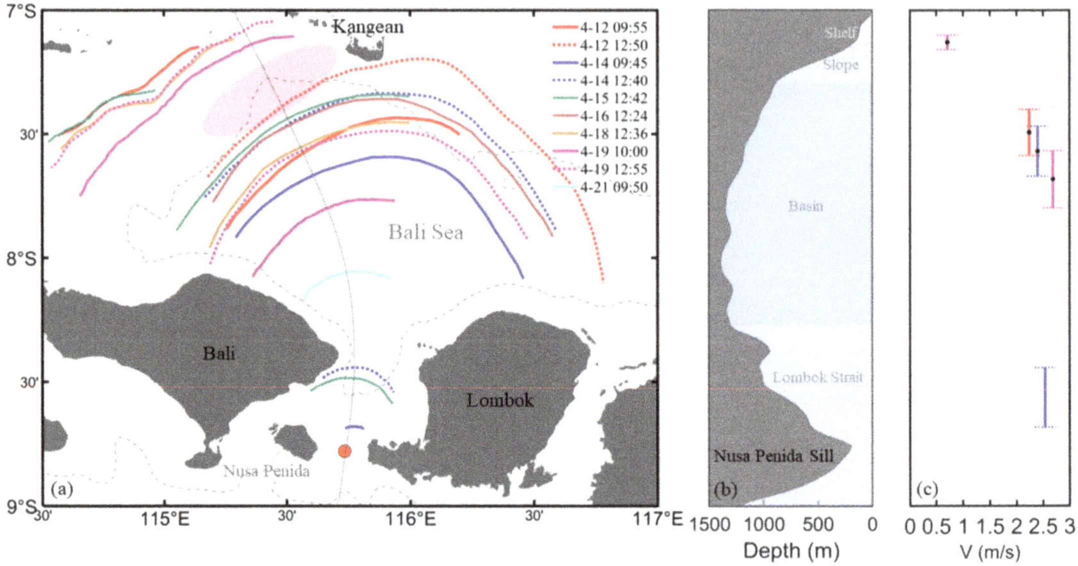

Figure 3. (**a**) The colored curves are crests of all leading ISWs in Figures 1 and 2. Gray dotted line is the 850 m isobath. The propagation path of ISWs and their generation site are plotted by thin gray line and a red dot, respectively. The pink ellipse marks the approximate area of the submarine wreck inferred from the reported information. (**b**) Bathymetry along the ISWs propagation path. (**c**) The average propagation speed between two points. Different colors correspond to ISWs in (**a**).

In the BS basin, there are three observed ISWs (S1, 4 and 5) whose speeds decreased from south to north along the propagation path. The images on 19 April (Figure 1e,f) showed that the wave crest center of S5 moved at a mean speed of 2.69 m/s between 7.80° S and 7.57° S where the average water depth was 1274 m, and further to the northwest of the basin, the images on 12 April (Figure 1a,b) suggested that the wave crest center of S1 propagated at a mean speed of 2.24 m/s between 7.59° S and 7.40° S. Over the shallow terrain west of the Kangean Islands, the propagation speed of S6 severely slowed down to a mean value of 0.71 m/s in the area with an average water depth of 155 m. The above calculations show a clear decreasing trend of propagation speed as ISWs shoaled from the deep basin of the BS onto the continental shelf. Through numerical simulation, Ningsih et al. [22] showed an ISW propagation speed range in the BS of 0.71–2.67 m/s, especially consistent with our estimated result. Using ALOS PALSAR images, Matthews et al. [21] defined ISW mean speeds between two wave packets about 1.6 to 2.3 m/s by measuring the distances between the leading signals in adjacent wave packets generated 12.4 h apart.

Figure 2a shows that S3 reached the continental shelf to the west of the Kangean Islands at 12:42 on April 15. Compared with the images taken on 14 April (Figure 1c,d), it took more than a day for the wave to travel from the LS to the continental shelf. During the over 27-h process from formation to shoaling in shallow water, the ISW traveled nearly 200 km and the average propagation speed was about 2 m/s.

3.3. Underwater Structure of ISWs Inferred from MODIS Images

Understanding the underwater structure of ISWs near the wreck site is crucial for evaluating the impacts of ISWs on submarine navigation. About two days before the submarine disaster, a clear wave packet (S5-b) near the wreck site was captured by the MODIS at 12:55 on 19 April (Figure 1f), and its underwater structure was reconstructed here.

According to the MODIS image, the center of the leading wave of S5 propagated in a northwest direction with a half-wavelength of ~2 km at a depth of about 1200 m (Figure 4a), and its amplitude was inversed to be 41 m. Furthermore, the theoretical propagation speed C of this ISW can be acquired by $C = C_0 + \alpha\eta_0/3$. In this case, C_0 and α are 2.51 m/s and −0.0166 respectively, and the calculated C is 2.73 m/s, which is in good agreement with the mean propagation obtained from the satellite images in Figure 1e,f.

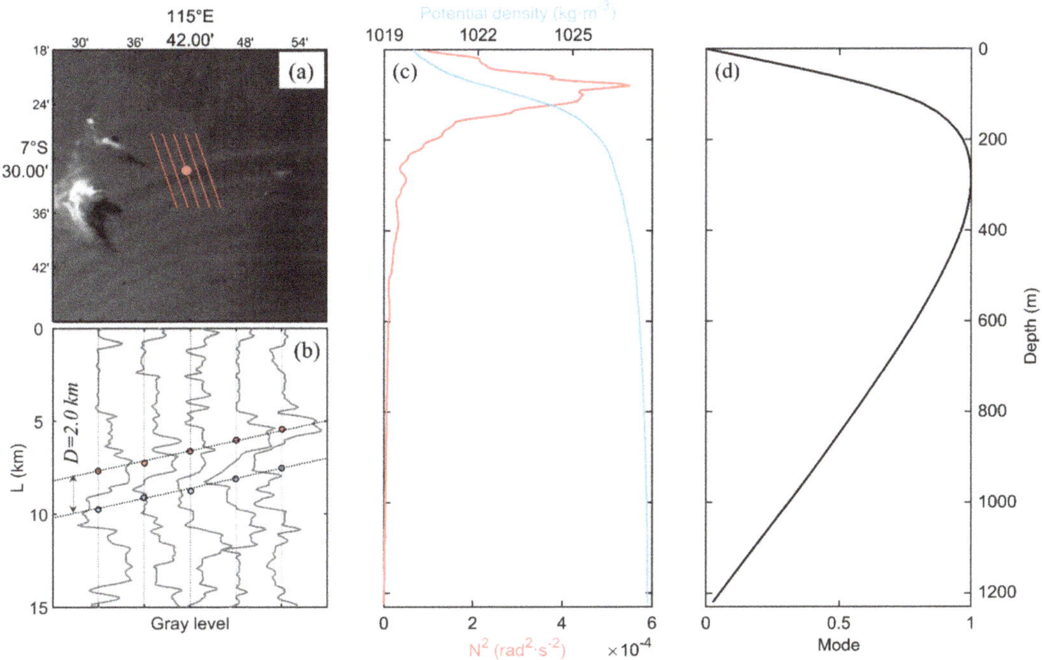

Figure 4. (**a**) Local amplification of the ISW occurring in the BS basin within the remote sensing image from 12:55 19 April local time. (**b**) Gray level changes along the red line in (**a**). (**c**) Climatological potential density and Brunt-Väisälä frequency profiles calculated from WOA18 data at the pink dot site in (**a**). (**d**) First-mode of the ISW.

Figure 5 shows the underwater structure of the wave S5-b. As shown in the figure, the maximum horizontal current velocity induced by the wave was 65 cm/s, and the flow core with a velocity greater than 50 cm/s existed in the upper 50 m and spent nearly 10 min passing by the site where the ISW crest located. Below 300 m, the horizontal current direction of the ISW was opposite to the wave propagation direction. Vertically, there were downward and upward currents respectively before and after the wave trough with a maximum velocity of 10 cm/s, and the vertical flow exceeding 6 cm/s extended for 800 m and 650 m in the horizontal and vertical directions, respectively.

3.4. Relationship between Barotropic Tides at Source and ISWs

It is generally believed that ISWs in the LS are generated by the interaction between tidal current and the Nusa Penida sill [17,23,31]. Understanding the relationship between the ISW generation and barotropic tides is of great significance for estimating the occurrence time of ISWs in the BS around the time of submarine wreck.

In the MODIS image taken at 9:45 on 14 April, it can be seen that the crest of S3-a was only about 10 km to the north of the source site (Figure 6a). Moreover, we calculated the internal Froude number (Fr) to examine the criticality of the tidal flow over the Nusa Penida sill. The Froude number (Fr = u/c) is a dimensionless quantity that expresses the ratio of barotropic flow speed (u) to the phase speed (c) of long internal wave [32]. At the above tidal peak, the Fr over the sill is 1.25 greater than 1, from which we can infer that S3 was likely released within the southward tidal. Backtracking at an average C_0 of 1.8 m/s, calculated from Equation (2), the generation time of S3 at the source site was about 08:12 (dashed red line in Figure 6b), adjacent to a TPXO southward tidal current peak. Purwandana et al. proposed that the generation of ISWs in the LS was related to the lee-wave mechanism [19], consistent with the near-source emergence of ISWs in the MODIS image in Figure 6a and the estimated release time of ISWs around the southward tidal current peak in Figure 6b.

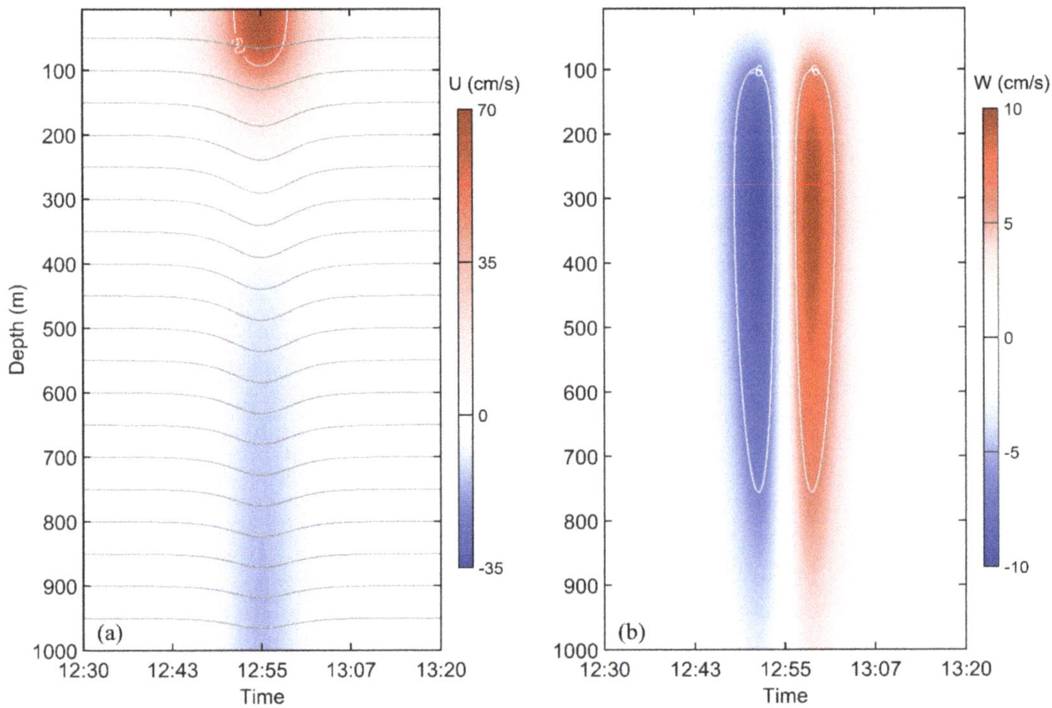

Figure 5. Underwater structure of the ISW. (**a**) Horizontal current along the wave propagation direction. The gray lines are isopycnal. (**b**) Vertical current. The white lines represent 6 cm/s isoline.

As previous measurements have revealed [19], the barotropic tidal current at the Nusa Penida sill was dominated by semi-diurnal component, and every day there were two southward barotropic tidal current peaks, occurring about 20–30 min later than the barotropic tides each day (Figure 6c). This indicates that two ISWs were generated in one day over the April tidal period. We traceback each ISW along the propagation path using the speed obtained from the satellite images and the calculated theoretical speed, and connect these ISWs, in Figure 6c and Table 1, with the suspected tidal peaks that generated them. From 11–20 April, the interval between the generation of two ISWs every day was reduced from 12.3 h to 11.7 h, slightly less than the 12.4 and the 12.42 h intervals defined by Matthews et al. and Karang et al. [21,29]. In addition, by comparing the distance between adjacent ISW crests and the time interval between their formation, we obtain a mean speed range of 1.35 (S1-a, S2) to 2.77 (S7, S8) m/s for the ISWs propagating in the BS.

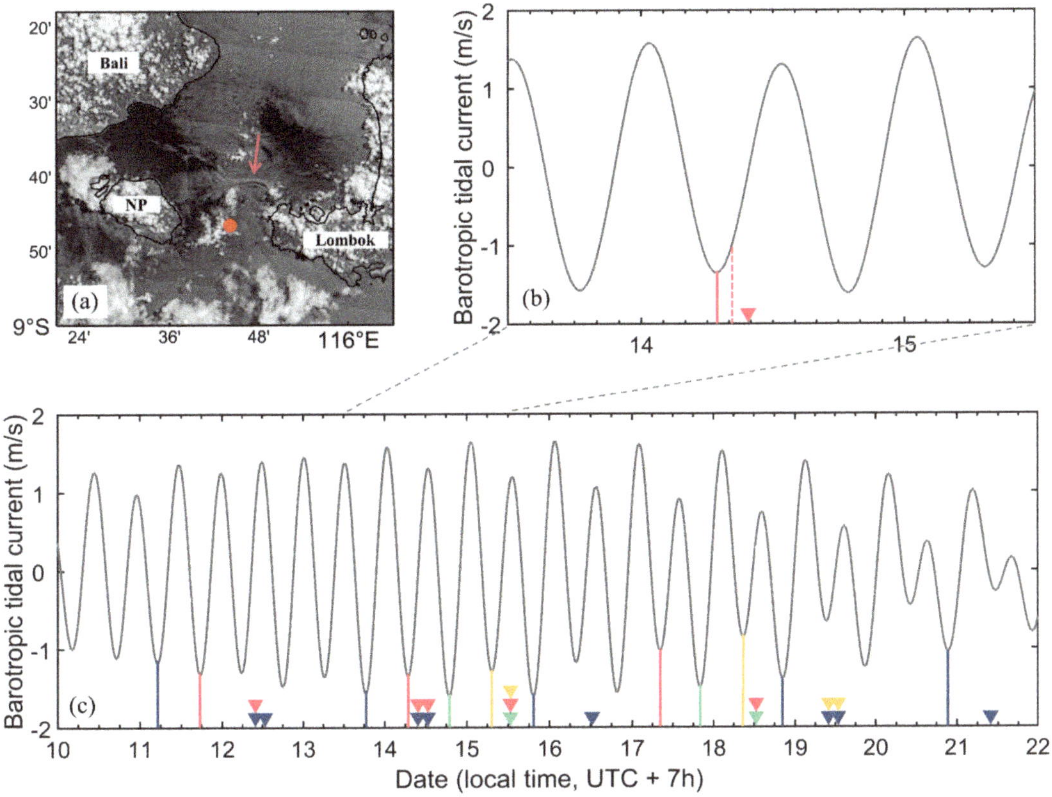

Figure 6. (**a**) Remote sensing images of S3 in the LS taken at 9:45 on 14 April local time. (**b**,**c**) Barotropic tide variation at the Nusa Penida sill from TPXO. The colored triangles represent the time when the ISWs were observed, and the dotted lines of the same color indicate the tidal current peaks associated with the generation of those ISWs.

Table 1. The observed time of ISWs in satellite images from 12 to 21 April 2021 and the occurrence time of southward tidal current peaks (STCP) associated with their generation.

ISW	Observed Time	STCP Time	ISW	Observed Time	STCP Time
S1	04–12 09:55 04–12 12:50	04–11 05:10	S7	04–15 12:42	04–15 07:20
S2	04–12 09:55	04–11 17:30	S8	04–15 12:42	04–14 19:00
S3	04–14 09:45 04–14 12:40 04–15 12:42	04–14 07:00	S9	04–16 12:24	04–15 09:20
S4	04–14 09:45 04–14 12:40	04–13 18:30	S10	04–18 12:36	04–17 20:00
S5	04–19 10:00 04–19 12:55	04–18 20:20	S11	04–18 12:36	04–17 08:20
S6	04–19 10:00 04–19 12:55	04–18 08:40	S12	04–21 19:50	04–20 21:20

4. Conclusions

By surveying remote sensing images over the BS, we found significant ISW activities near the submarine *KRI nanggala-402* wreck site in April 2021. Those ISWs were generated in the LS and then traveled along the north-western direction in the BS basin, passing through the submarine wreck site, and finally reached the continental shelf to the west of the Kangean Islands. The ISWs in April 2021 were mainly distributed in the area to the north of Bali Island, rather than north of Lombok Island as was revealed by previous studies, indicating a significant temporal variation of ISW distribution in the BS.

Along the propagation path, there were up to three wave packets simultaneously existing in the BS. The wave packet contained dozens of solitons, whose crest can extend for 200 km, within a meridional range of more than 60 km, covering a vast region of the BS. Those ISWs propagated at a mean speed of 2 m/s from the source region to the continental shelf, and the speed was as fast as 2.69 m/s in the BS basin and reduced to 0.71 m/s in the shallow water. On 19 April, about two days before the submarine incident, the amplitude of an ISW near the submarine wreck site was inversed to be 41 m according to satellite images, and the reconstructed underwater structure showed a maximum horizontal and vertical velocity of 65 cm/s and 10 cm/s, respectively. Moreover, it was inferred from the near-source evidence that ISWs were released within the southward barotropic tidal trough, and the variation of source tides revealed that two ISWs were generated with an interval of 11.7 to 12.3 h every day during the April tidal period.

The analyses presented here have provided necessary observational information on the ISW activities in the BS for the submarine wreck investigations, and whether or not the submarine *KRI nanggala-402* encountered with ISWs will be ascertained once the accurate time and location site of the submarine wreck becomes available in the future. In addition, it seemed that ISWs in the area to the north of Bali Island were extraordinarily active around the time of the submarine wreck in comparison with the statistical results from 2006 to 2011 and from 2014 to 2015 based on satellite images [14,29]. Such variability of ISW distribution is also an interesting topic worth further investigation, and long in-situ observations prove the necessary to improve the understanding of the ISWs in BS.

Author Contributions: Conceptualization, X.H., T.W., W.Z., S.Z., Y.Y. and J.T.; methodology, T.W. and X.H.; software, Y.Y.; validation, T.W. and X.H.; formal analysis, T.W.; investigation, T.W. and X.H.; resources, X.H. and W.Z.; data curation, T.W.; writing—original draft preparation, T.W.; writing—review and editing, X.H., W.Z., S.Z. and Y.Y.; visualization, T.W.; supervision, X.H.; project administration, W.Z.; funding acquisition, X.H., W.Z. and J.T. All authors have read and agreed to the published version of the manuscript.

Funding: This paper was supported by the National Natural Science Foundation of China (Grants 41976008, 91858203, 91958205), the Hainan Provincial Joint Project of Sanya Yazhou Bay Science and Technology City (Grant 120LH018).

Data Availability Statement: The data presented in this study are available on request from the corresponding author.

Acknowledgments: We acknowledge the use of satellite images from the NASA Worldview application (https://worldview.earthdata.nasa.gov, accessed on 28 November 2021). The WOA18 data was downloaded from the NOAA's National Centers for Environmental Information (https://www.ncei.noaa.gov/products/world-ocean-atlas, accessed on 25 November 2021).

Conflicts of Interest: The authors declare no conflict of interest.

References

1. Alford, M.H.; Mickett, J.B.; Zhang, S.; MacCready, P.; Zhao, Z.; Newton, J. Internal Waves on the Washington Continental Shelf. *Oceanography* **2012**, 25, 66–79. [CrossRef]
2. Da Silva, J.C.B.; New, A.L.; Srokosz, M.A.; Smyth, T. On the observability of internal tidal waves in remotely-sensed ocean colour data. *Geophys. Res. Lett.* **2002**, 29, 10-1–10-4. [CrossRef]
3. Huang, X.; Wang, Z.; Zhang, Z.; Yang, Y.; Zhou, C.; Yang, Q.; Zhao, W.; Tian, J. Role of Mesoscale Eddies in Modulating the Semidiurnal Internal Tide: Observation Results in the Northern South China Sea. *J. Phys. Oceanogr.* **2018**, 48, 1749–1770. [CrossRef]

4. Huang, X.; Zhang, Z.; Zhang, X.; Qian, H.; Zhao, W.; Tian, J. Impacts of a Mesoscale Eddy Pair on Internal Solitary Waves in the Northern South China Sea revealed by Mooring Array Observations. *J. Phys. Oceanogr.* **2017**, *47*, 1539–1554. [CrossRef]
5. Kinder, T.H. Net mass transport by internal waves near the Strait of Gibraltar. *Geophys. Res. Lett.* **1984**, *11*, 987–990. [CrossRef]
6. Raju, N.J.; Dash, M.K.; Dey, S.P.; Bhaskaran, P.K. Potential generation sites of internal solitary waves and their propagation characteristics in the Andaman Sea—a study based on MODIS true-colour and SAR observations. *Environ. Monit. Assess.* **2019**, *191*, 809. [CrossRef] [PubMed]
7. Ramp, S.; Tang, T.Y.; Duda, T.; Lynch, J.; Liu, A.; Chiu, C.-S.; Bahr, F.; Kim, H.-R.; Yang, Y.-J. Internal Solitons in the Northeastern South China Sea Part I: Sources and Deep Water Propagation. *IEEE J. Ocean. Eng.* **2004**, *29*, 1157–1181. [CrossRef]
8. Zhang, X.; Huang, X.; Zhang, Z.; Zhou, C.; Tian, J.; Zhao, W. Polarity Variations of Internal Solitary Waves over the Continental Shelf of the Northern South China Sea: Impacts of Seasonal Stratification, Mesoscale Eddies, and Internal Tides. *J. Phys. Oceanogr.* **2018**, *48*, 1349–1365. [CrossRef]
9. Huang, X.; Chen, Z.; Zhao, W.; Zhang, Z.; Zhou, C.; Yang, Q.; Tian, J. An extreme internal solitary wave event observed in the northern South China Sea. *Sci. Rep.* **2016**, *6*, 30041. [CrossRef]
10. Osborne, A.; Burch, T.; Scarlet, R. The Influence of Internal Waves on Deep-Water Drilling. *J. Pet. Technol.* **1978**, *30*, 1497–1504. [CrossRef]
11. Chen, J.; You, Y.X.; Liu, X.D.; Wu, C.S. Numerical simulation of interaction of internal solitary waves with a moving submarine. *J. Hydrodyn.* **2010**, *25*, 344–351. (In Chinese)
12. Huang, M.M.; Zhang, N.; Zhu, A.J. Hydrodynamic loads and motion features of a submarine with interaction of internal solitary waves. *J. Ship Mech.* **2019**, *23*, 531–540. (In Chinese)
13. Stepanyants, Y. How internal waves could lead to wreck American and Indonesian submarines? *arXiv* **2021**, arXiv:2107.00828.
14. Karang, I.W.G.A.; Nishio, F.; Mitnik, L.; Osawa, T. Spatial-Temporal Distribution and Characteristics of Internal Waves in the Lombok Strait Area Studied by Alos-Palsar Images. *Earth Sci. Res.* **2012**, *1*, 11. [CrossRef]
15. Matthews, J.; Awaji, T. Synoptic mapping of internal-wave motions and surface currents near the Lombok Strait using the Along-Track Stereo Sun Glitter technique. *Remote Sens. Environ.* **2010**, *114*, 1765–1776. [CrossRef]
16. Mitnik, L.; Alpers, W.; Chen, K.S.; Chen, A.J. Manifestation of internal solitary waves on ERS SAR and SPOT images: Similarities and differences. In Proceedings of the IGARSS 2000. IEEE 2000 International Geoscience and Remote Sensing Symposium. Taking the Pulse of the Planet: The Role of Remote Sensing in Managing the Environment, Honolulu, HI, USA, 24–28 July 2000; Volume 5, pp. 1857–1859. [CrossRef]
17. Susanto, R.D.; Mitnik, L.; Zheng, Q. Ocean Internal Waves Observed in the Lombok Strait. *Oceanography* **2005**, *18*, 80–87. [CrossRef]
18. Syamsudin, F.; Taniguchi, N.; Zhang, C.; Hanifa, A.D.; Li, G.; Chen, M.; Mutsuda, H.; Zhu, Z.; Zhu, X.; Nagai, T.; et al. Observing Internal Solitary Waves in the Lombok Strait by Coastal Acoustic Tomography. *Geophys. Res. Lett.* **2019**, *46*, 10475–10483. [CrossRef]
19. Purwandana, A.; Cuypers, Y.; Bouruet-Aubertot, P. Observation of internal tides, nonlinear internal waves and mixing in the Lombok Strait, Indonesia. *Cont. Shelf Res.* **2021**, *216*, 104358. [CrossRef]
20. Aiki, H.; Matthews, J.P.; Lamb, K.G. Modeling and energetics of tidally generated wave trains in the Lombok Strait: Impact of the Indonesian Throughflow. *J. Geophys. Res. Earth Surf.* **2011**, *116*, C03023. [CrossRef]
21. Matthews, J.P.; Aiki, H.; Masuda, S.; Awaji, T.; Ishikawa, Y. Monsoon regulation of Lombok Strait internal waves. *J. Geophys. Res. Earth Surf.* **2011**, *116*, C05007. [CrossRef]
22. Ningsih, N.S.; Rachmayani, R.; Hadi, S.; Brodjonegoro, I.S. Internal Waves Dynamics in The Lombok Strait Studied By A Numerical Model. *Int. J. Remote Sens. Earth Sci.* **2010**, *5*, 17–33. [CrossRef]
23. Mitnik, L.; Alpers, W.; Lim, H. Thermal plumes and internal solitary waves generated in the Lombok Strait studied by ERS SAR. In Proceedings of the ERS-Envisat Symposium: Looking down to Earth in the New Millennium, Gothenburg, Sweden, 16–20 October 2000.
24. Zhao, Z.X.; Klemas, V.V.; Zheng, Q.N.; Yan, X.H. Satellite observation of internal solitary waves converting polarity. *Geophys. Res. Lett.* **2003**, *30*, 1988. [CrossRef]
25. Jackson, C. Internal wave detection using the Moderate Resolution Imaging Spectroradiometer (MODIS). *J. Geophys. Res. Earth Surf.* **2007**, *112*, C11012. [CrossRef]
26. Huang, X.D.; Zhao, W. Information of Internal Solitary Wave Extracted from MODIS image: A Case in the Deep Water of Northern South China Sea. *Period. Ocean. Univ. China* **2014**, *44*, 19–23.
27. Ning, J.; Wang, J.; Zhang, M.; Cui, H.J.; Lu, K.X. Amplitude Inversion Model and Application of Internal Solitary Waves of the Northern South China Sea Based on Optical Remote-sensing Images. *ACTA Photonica Sinica* **2019**, *48*, 1228003. (In Chinese) [CrossRef]
28. Zheng, Q.; Yuan, Y.; Klemas, V.; Yan, X.-H. Theoretical expression for an ocean internal soliton synthetic aperture radar image and determination of the soliton characteristic half width. *J. Geophys. Res. Earth Surf.* **2001**, *106*, 31415–31423. [CrossRef]
29. Karang, I.W.G.A.; Chonnaniyah, C.; Osawa, T. Landsat 8 Observation of the Internal Solitary Waves in the Lombok Strait. *Indones. J. Geogr.* **2019**, *51*, 251–260. [CrossRef]

30. Lindsey, D.T.; Nam, S.; Miller, S.D. Tracking oceanic nonlinear internal waves in the Indonesian seas from geostationary orbit. *Remote Sens. Environ.* **2018**, *208*, 202–209. [CrossRef]
31. Mitnik, L.; Alpers, W. Sea surface circulation through the Lombok Strait studied by ERS SAR. In Proceedings of the 5th Pacific Ocean Remote Sensing Conference (PORSEC 2000), Goa, India, 5–8 December 2000; Volume I, pp. 313–317.
32. Farmer, D.M.; Smith, J.D. Tidal interaction of stratified flow with a sill in Knight Inlet. *Deep Sea Res. Part A Oceanogr. Res. Pap.* **1980**, *27*, 239–254. [CrossRef]

Journal of
Marine Science and Engineering

MDPI

Article

The Impact of Fortnightly Stratification Variability on the Generation of Baroclinic Tides in the Luzon Strait

Zheen Zhang [1,*], Xueen Chen [1] and Thomas Pohlmann [2]

[1] College of Oceanic and Atmospheric Sciences, Ocean University of China, Qingdao 266100, China; xchen@ouc.edu.cn
[2] Centre for Earth System Research and Sustainability, Institute of Oceanography, University of Hamburg, 20146 Hamburg, Germany; thomas.pohlmann@uni-hamburg.de
* Correspondence: zhangzheen@live.com

Abstract: The impact of fortnightly stratification variability induced by tide–topography interaction on the generation of baroclinic tides in the Luzon Strait is numerically investigated using the MIT general circulation model. The simulation shows that advection of buoyancy by baroclinic flows results in daily oscillations and a fortnightly variability in the stratification at the main generation site of internal tides. As the stratification for the whole Luzon Strait is periodically redistributed by these flows, the energy analysis indicates that the fortnightly stratification variability can significantly affect the energy transfer between barotropic and baroclinic tides. Due to this effect on stratification variability by the baroclinic flows, the phases of baroclinic potential energy variability do not match the phase of barotropic forcing in the fortnight time scale. This phenomenon leads to the fact that the maximum baroclinic tides may not be generated during the maximum barotropic forcing. Therefore, a significant impact of stratification variability on the generation of baroclinic tides is demonstrated by our modeling study, which suggests a lead–lag relation between barotropic tidal forcing and maximum baroclinic response in the Luzon Strait within the fortnightly tidal cycle.

Keywords: Luzon Strait; baroclinic tides; stratification variability; MITgcm

Citation: Zhang, Z.; Chen, X.; Pohlmann, T. The Impact of Fortnightly Stratification Variability on the Generation of Baroclinic Tides in the Luzon Strait. *J. Mar. Sci. Eng.* **2021**, *9*, 703. https://doi.org/10.3390/jmse9070703

Academic Editors: Déborah Idier and Vengatesan Venugopal

Received: 30 March 2021
Accepted: 24 June 2021
Published: 26 June 2021

1. Introduction

Internal waves for which reduced gravity provides the restoring force widely occur in seas and are responsible for enhanced dissipation in the deep ocean [1]. Most of the internal wave energy is radiated from the source over a long distance, which can significantly affect the space-time distribution of energy available for global mixing [2–4]. The South China Sea may have most intense internal waves among all oceans. Internal solitary waves with peak-to-trough amplitudes over 150 meters have been reported in the northern South China Sea, and such waves are believed to impact other ocean processes notably [5,6]. Both observation and numerical simulation studies have suggested that the Luzon Strait is the main generation site of internal waves in the northern South China Sea [7–14]. Alford et al. depicted a cradle-to-grave picture of internal waves from the Luzon Strait to the continental shelf on a basin scale through the combination of in situ data and numerical simulation [15]. In situ observations show that the regularity and strength of internal solitary waves on the shelf of the northern South China Sea has a significant spring-neap variability [12,15,16]. Moreover, internal tides in the deep basin west of the Luzon Strait also show a spring-neap variability that has been demonstrated by moored current observations [17]. In addition, numerical simulations and remote sensing data suggest that internal solitary waves on the shelf are developed by nonlinear steepening and frequency dispersion of the baroclinic tides generated in the Luzon Strait by tide–topography interactions [8,9,12,15,18]. As Ramp et al. stated, "Most of the features of the wave arrival can be explained by the tidal variability in the Luzon Strait" [16]. Therefore, the generation of baroclinic tides in the Luzon Strait, which is investigated in the present study, is important for forecasting internal wave characteristics in the South China Sea.

At least five different internal wave generation mechanisms exist including beam scattering, mixed region collapse, and lee–wave mechanism [18–22]. The timing and strength of the energy transfer process from barotropic to baroclinic mode thereof determine how the internal wave is generated at the beginning. In previous numerical studies concerning baroclinic tides in the Luzon Strait, the main focus has been on tides and topography and their interaction, for example, the spring-neap cycle, flood–ebb current, generation site, and resonance between two ridges [11–13,23,24]. Stratification is considered a minor factor in internal wave variability, for example, the seasonal variation of internal tides in the northern South China Sea is considered to be mainly modulated by the astronomical tides rather than by the seasonal thermocline in the Luzon Strait [17]. Some previous numerical simulations have shown that stratification can notably impact internal wave generation under specific conditions [25,26]. Recently, an extreme internal solitary wave with a maximum peak-to-though amplitude of 240 m was reported in the northern South China Sea [6]. The authors argue that the variability of stratification in the Luzon Strait may be a key factor for the generation of energetic internal tides and the formation of this extreme internal solitary wave event. As a result, the variability of stratification in the Luzon Strait obviously deserves more attention in baroclinic tide generation studies. In particular, stratification can be affected by many factors, such as surface heat flux and mesoscale eddy intrusion. To simplify this situation, we focus only on the tidal effect on stratification because the interaction between strong tidal flow and steep topography in the Luzon Strait can severely affect the local stratification and is thus the major source of baroclinic variability in our case. The specific questions we address are how stratification is affected by tide–topography interactions and how the baroclinic tides are affected by this stratification variability. Research on these questions will be helpful for improving the understanding of the internal wave generation process and variability in realistic situations.

In this paper, the MIT general circulation model (MITgcm) described by Marshall et al. is used for the three-dimensional hydrodynamic baroclinic simulation [27]. We focus mainly on the stratification variability and the generation of baroclinic tides in the Luzon Strait. Therefore, we decided to use the hydrostatic version of MITgcm as our main target processes can be adequately resolved by this model. The main objective is to determine how the stratification variability affects the internal wave generation in the Luzon Strait.

In Section 2, the model settings are presented. Subsequently, in Section 3, the model validation, the stratification variability, and the analysis of governing mechanisms in the Luzon Strait are presented. These sections are followed by an analysis of energy transfer and its effect on internal wave generation. Finally, the discussion and conclusions are presented in Sections 4 and 5, respectively.

2. Materials and Methods

The model domain covers the northern South China Sea, the Luzon Strait, and part of the Western Pacific and ranges zonally from 110° E to 129° E and meridionally from 16° N to 23° N. Realistic bathymetry data are derived from the general bathymetric chart of the oceans (GEBCO_08, Figure 1). The primary research domain, i.e., the Luzon Strait, is placed at the center of the model domain. L1 is selected as a generation site, and B2 falls along the propagation path [16]. The grid resolution in the horizontal direction is set to 1/30th degree × 1/30th degree, which is fine enough to describe baroclinic tidal signals for our model domain. A total of 115 uneven vertical layers are set, ranging from 10 m in the upper ocean and gradually increasing to 200 m in the deep ocean.

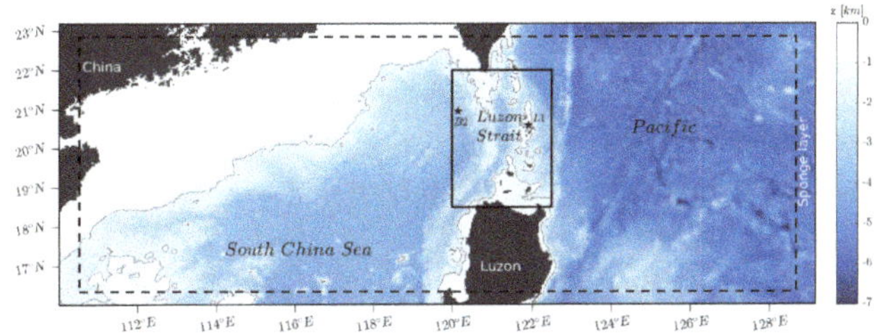

Figure 1. Model domain and its topography. Black solid box is the main research domain, the Luzon Strait. Selected locations L1 and B2, which refer to Ramp et al., are marked with stars [16]. The dashed line is the inner boundary of a sponge layer.

The model is driven by barotropic tidal currents composed of the first eight tidal constituents (M_2, S_2, K_2, N_2, K_1, O_1, P_1, Q_1) at four open lateral boundaries. Harmonic constants of forcing are derived from the OSU TOPEX/Poseidon Global Inverse Solution (TPXO7.2). A sponge layer is specified along the open boundaries to avoid artificial reflection (Figure 1). A relaxation term that relaxes variables toward the boundary values by applying a linearly increasing relaxation time scale is added to the momentum equation. The relaxation time scale is set to one M_2 cycle at the interior termination of the sponge layer and to one thousandth of one M_2 cycle at boundaries as the M_2 signal is most significant in our case.

Initial temperature and salinity are derived from the World Ocean Atlas 2009 and are horizontally homogeneous and vertically stratified (Figure 2a,b). According to the buoyancy frequency N (Figure 2c), the initial pycnocline is at a depth of approximately 100 m, showing a stable ocean stratification structure. A no-slip condition is used for lateral boundaries and at the bottom. The quadratic bottom drag coefficient is 0.002. The vertical diffusivity and viscosity are calculated by the KPP vertical mixing parameterizations proposed by Large et al., which can simulate oceanic processes like convective penetration and diurnal cycling based on physical principles [28]. The model run lasts for 50 model days with 1-min time steps and hourly output.

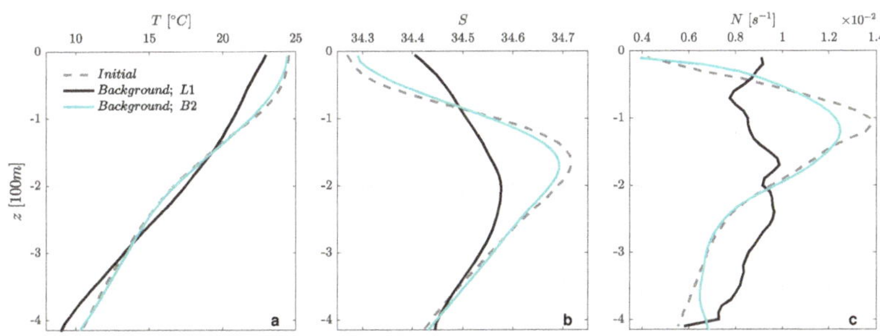

Figure 2. (**a**) The temperature, (**b**) salinity, and (**c**) buoyancy frequency profiles. Gray dashed lines are the initial field. Black and blue solid lines are the background profiles at L1 and B2, respectively.

3. Results

3.1. Model Validation

Considering that we only use barotropic tidal forcing to drive the model, the model run is primarily validated using TPXO7.2 data. Figure 3 gives the co-tidal charts for the first four tidal constituents derived through harmonic analysis over the final 30 days of the model results. The co-tidal chart of M_2 (Figure 3a) is in good agreement with previous tidal current measurements [29]. Our model also reproduced the degenerated counterclockwise amphidromic system of K_1 and O_1 near the northern tip of Luzon Island, which was mentioned in previous simulations [23,30,31]. Notably, wave-like stripes exist in the model-predicted amplitude and phase. These features are the manifestation of spatially coherent internal tide modulations with associated wavelengths [32].

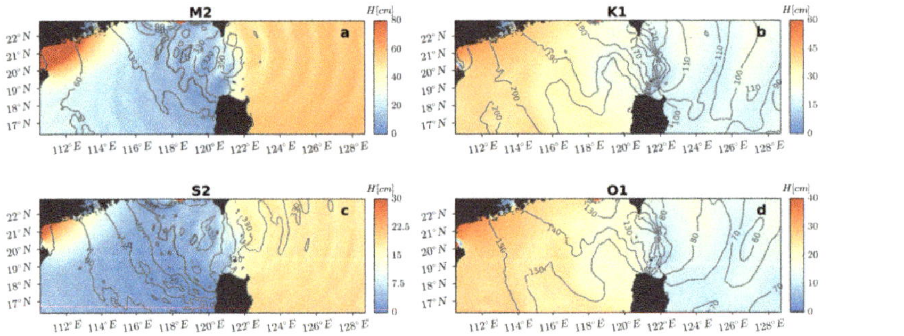

Figure 3. The co-tidal charts calculated from model-predicted sea-level height for (**a**) M_2, (**b**) K_1, (**c**) S_2, and (**d**) O_1. Patched colors represent amplitude, gray lines represent phase (in deg and in reference to 0°).

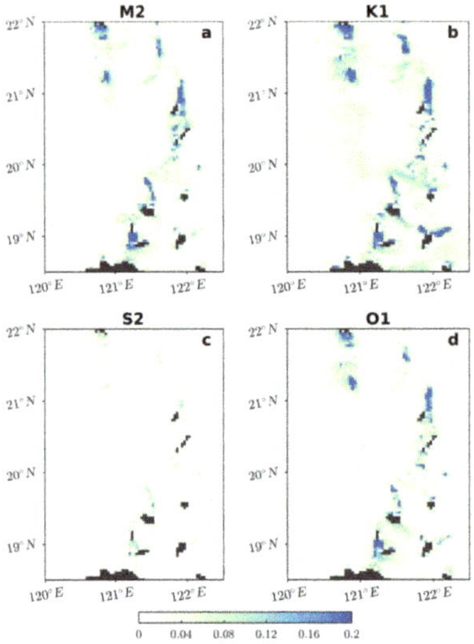

Figure 4. The root mean square difference between the model and TPXO7.2 data of zonal barotropic velocity for (**a**) M_2, (**b**) K_1, (**c**) S_2, and (**d**) O_1.

The root mean square difference D between the model and TPXO7.2 data for sea-level height and zonal barotropic velocity is used here [33]. The barotropic velocities $U = \frac{1}{H}\overline{u}$ and $V = \frac{1}{H}\overline{v}$ are defined as depth-averaged velocities, where $\overline{(\cdot)} = \int_{-d}^{\eta}(\cdot)dz$ is the depth integration of a quantity from the bottom $-d$ to surface η and the total water depth $H = \eta + d$.

$$D = \sqrt{\frac{1}{2}(A_o^2 + A_m^2) - A_o A_m \cos(\phi_o - \phi_m)} \quad (1)$$

where A and ϕ are the amplitudes and phases of a given constituent, respectively, and subscripts $_o$ and $_m$ refer to the TPXO7.2 data and our model, respectively. The domain-averaged D of sea-level height in areas deeper than 200 m is 3.03, 5.46, 1.34, and 2.78 cm for M_2, K_1, O_1, and S_2, respectively. The distributions of D of the zonal barotropic velocity in the Luzon Strait for the first four tidal constituents (Figure 4) show small values in most regions, which indicates a good agreement between the model results and the TPXO7.2. Relatively larger D values appear mainly above ridges, where water is shallow. Our model is fully baroclinic and has a slightly different topography compared to the TPXO7.2 model, which has a stronger effect in shallow areas. Data assimilation is not included in our case but may also result in differences.

Figure 5 shows the comparison of time series of sea-level height η at L1 between our model prediction and TPXO7.2. L1 is located on the side of a seamount in the Luzon Strait and is hypothesized to be a generation source for internal waves [16]. These two time series are in good agreement, and both show well-recognized spring-neap characteristics. The heights of high tides and low tides are different, but the phases match very well. Therefore, our model is suitable for the simulation of tidal characteristics.

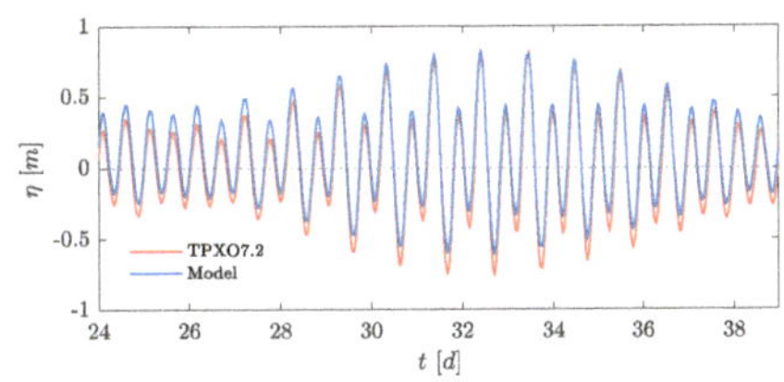

Figure 5. The time series of sea-level height at L1. The blue line represents the model results, the red line represents the TPXO7.2 results.

3.2. Stratification Variability

The variability of stratification in our investigation is an important part of baroclinic responses. Therefore, we investigate this feature in more detail. The depth and strength of the pycnocline are two parameters used to evaluate stratification, whereas buoyancy frequency N provides a more concise measure of ocean stratification. Given that barotropic tidal current in the Luzon Strait has a significant spring-neap tidal cycle, we calculate a time average over the spring-neap tidal cycle (model days 24 to 38) to determine the background field. For example, the background buoyancy frequency $N_b = \langle N \rangle$, where $\langle \cdot \rangle = \frac{1}{T}\int_t^{t+T}(\cdot)d\tau$, is the time average of a quantity over the time interval T, T is 15 days covering a spring-neap cycle in this case. Apparently, the model results show different background stratification at different locations (Figure 2). At L1, which is located on the eastern side of the ridge and has a sharp topography, the N_b shows no prominent peaks, which means the water column here has experienced significant vertical mixing and diffusion since initialization. At B2, which is located in the sea basin with more than 3000 m depth, N_b is similar to the initial state, which means that stratification here is only slightly changed and that the vertical structure of the water column can be well maintained. For a more integrated picture, we analyze the time series of the depth-averaged N^2 between the surface and a depth of 300 m instead of the profiles below (Figure 6a).

According to the time series of the depth-averaged N^2, the stratification at L1 exhibits daily oscillations. After applying two times a moving average with a window size of 24 h, the time-smoothed result of stratification is marked by a bold line indicating the approximate fortnightly variability (Figure 6a). This variability shows a buoyancy increase and decrease cycle at L1 in the model, although no buoyancy input takes place during the model run, leading to the question as to which process is responsible for the enhancement of stratification. The low-pass decline of the depth-averaged N^2 is due to numerical diffusion and mixing. Finally, the stratification will almost disappear according to our model setting of no surface buoyancy forcing. The boundary tidal current signal plays an important role on the stratification variability, as it quickly traverses the model domain and dominates its variability. To investigate the reason for these results, we reviewed the local barotropic current field first (Figure 6b,c). The time series of zonal and meridional barotropic velocities suggest that the local barotropic flows are asymmetric, which means that these flows have different strengths in opposite directions, and the stratification shows a similar spring-neap cycle as the barotropic velocity, which means there is an impact from barotropic forcing on stratification. The asymmetry in the barotropic tidal flows is defined by the discrepancies in the duration of the eastward (northward) and westward (southward) tidal currents [18]. The interactions and phase difference between tidal constituents are the major source behind the barotropic tidal asymmetry [34,35]. Besides, there are some barotropic mean flows on the order of 0.1 m/s near the west ridge (not shown), which are considered to be caused by topographic rectification [36–38], also contribute to the asymmetric barotropic flows.

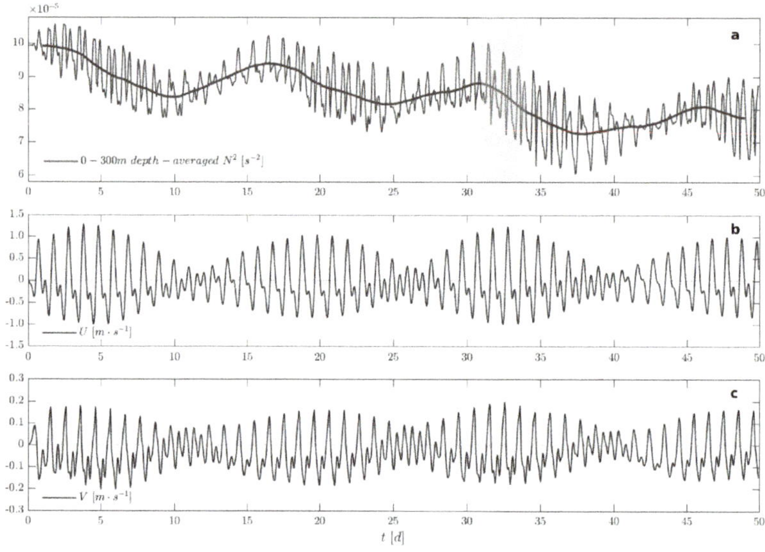

Figure 6. (**a**) The time series of 0–300-meter depth-averaged N^2 and corresponding time-smoothed result (bold line) at L1. The shaded bar represents a three-day period. (**b**) The zonal barotropic U and (**c**) V at L1.

In the following, we introduce the potential energy anomaly $\varphi = \frac{g}{H}\overline{(\hat{\rho} - \rho)z}$ [39,40] to explain what happens during stratification changes, where g is the gravity acceleration and $\hat{\rho}$ is the depth-averaged density. φ is a depth-integrated value that represents changes in potential energy relative to the vertically homogeneous conditions. For a given density profile, φ is the amount of work per unit volume required to completely homogenize the water column [40]. Thus, we define $\Delta\varphi = \varphi - \varphi_0$ as the potential energy anomaly change, where φ_0 is calculated from the initial field.

The interaction between barotropic flows and ridges causes vertical movements (Figure 7a) and thus can uplift or depress the isopycnals. Through this physical process, φ can decrease or increase (Figure 7b). Considering the generation of internal waves here and the accompanying intense baroclinic currents, the baroclinic component is expected to dominate the vertical velocity. These baroclinic flows are associated with the asymmetric barotropic forcing and exhibit a fortnightly cycle. Thus it is evident that the accumulated enhancing of stratification is unequal to the weakening (Figure 7b), and finally overwhelms the φ, thereby contributing to the fortnightly stratification variability (Figure 6a).

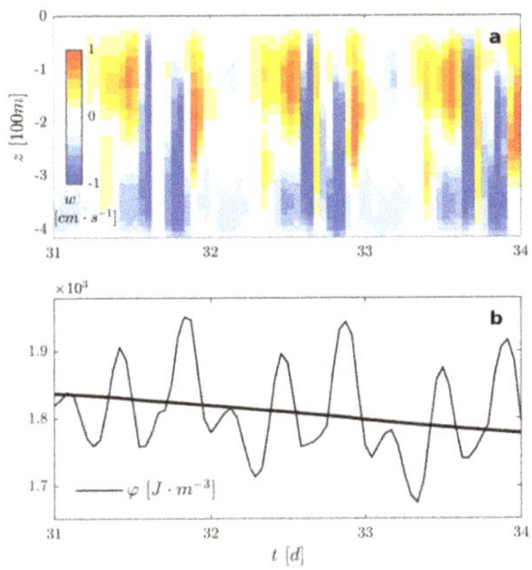

Figure 7. (**a**) The vertical velocity w (positive upward) and (**b**) the potential energy anomaly φ and the corresponding time-smoothed result at L1 over three days (shaded in Figure 6a).

Figure 8 shows the position of the isopycnals in relation to the baroclinic flow in the zonal L1-section during a tidal cycle. The interaction between barotropic forcing and topography in the stratified ocean can produce intense baroclinic currents. The simulated baroclinic currents show a structure of wave beams and their speed can reach to about 0.5 m/s, which are in agreement with the in situ observations [38]. Therefore, it can be concluded that our model performance is acceptable. According to our simulation, the horizontal baroclinic currents change directions with the tidal phase. The isopycnals also vary during a tidal cycle. In addition, the horizontal baroclinic velocity at L1 beats at a fortnightly cycle (not shown), which implies the impact of horizontal bariclinic currents on fortnightly stratification variability.

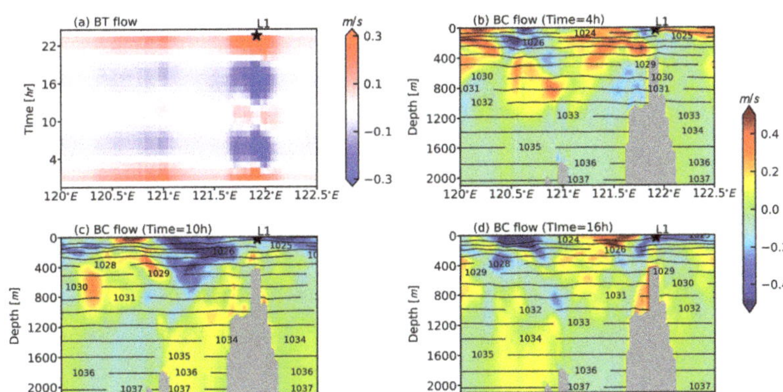

Figure 8. (**a**) The daily cycle of barotropic U-velocity in the zonal L1-section of day 25. The corresponding baroclinic U-velocity (shaded) and the isopycnals (contours) at (**b**) 4:00, (**c**) 10:00, and (**d**) 16:00.

Figure 9 shows daily averages of temperature and salinity when the time-smoothed depth-averaged N^2 is ascending or descending. Daily averages of the baroclinic U-velocity and V-velocity as well as the full W-velocity are shown in Figure 10. As can be seen, the magnitude of these mean currents and their temporal variability are in a reasonable range, indicating that the model is able to reasonably reproduce the underlying processes. The baroclinic velocity fields are associated with the asymmetric barotropic forcing in our simulation, and thus these rectified baroclinic flows also exhibit a fortnightly variability.

As shown in Figure 9, the halocline at L1 is depressed at day 25 and it is uplifted at day 32 above the ridge. Considering the vertical rectified flow exists in this area (Figure 10), the vertical movement of the halocline demonstrates that the vertical advection of buoyancy by the rectified flow near the ridge contributes to the fortnightly stratification variability. Due to the effect of internal wave generation, the stratification at L1 is relatively weak compared to its surroundings. Considering the horizontal difference of stratification and the existence of the horizontal mean baroclinic flow in this area (Figure 10), the horizontal advection by the baroclinic flow also contributes to the fortnightly stratification variability.

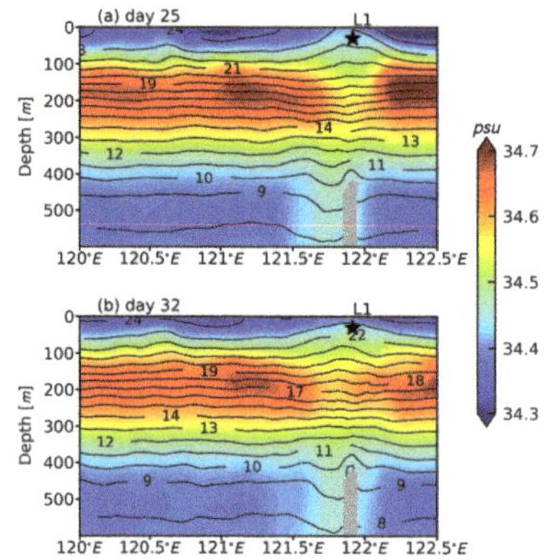

Figure 9. Daily averages of salinity and temperature in the zonal L1-section over (**a**) the ascending part day 25 and (**b**) the descending part day 32 shown in Figure 6.

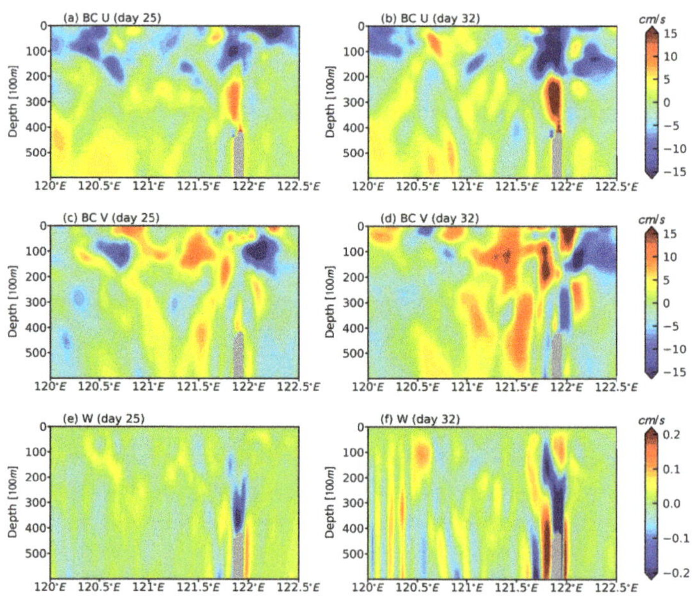

Figure 10. Daily averages of (**a**,**b**) baroclinic U-velocity, (**c**,**d**) baroclinic V-velocity, and (**e**,**f**) W-velocity in the zonal L1-section over the ascending part day 25 and the descending part day 32.

These dynamics can explain how the stratification process varies at L1. Subject to our model configuration, the energy that enhances the stratification must originally come from boundary forcing. Given that there are two ridges in the Luzon Strait, when the stratification at L1 is enhanced, the stratification at other places within the Luzon Strait should be weakened. Figure 11a,b shows $\Delta\varphi$ and the corresponding time-smoothed result at L1 and L2, respectively, where L1 is on the east side of the ridge and L2 is on the west side of the ridge. The time-smoothed result at L2 shows an almost inverse phase compared to that at L1, which proves our speculation. In order to determine the spatial distribution, we calculated three daily averages $\langle\Delta\varphi\rangle$ that are separated by 6 days, $d1$,$d2$ and $d3$ (Figure 11a).

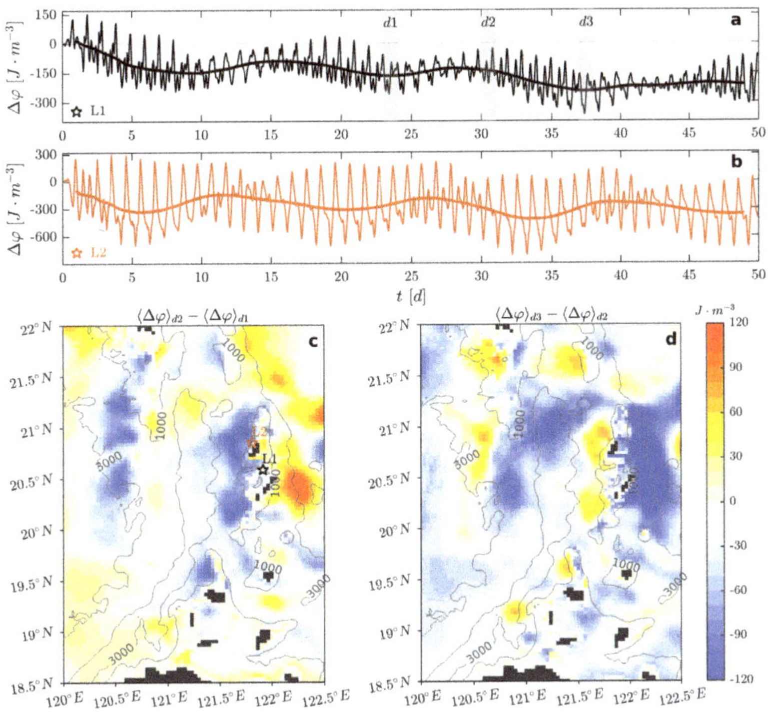

Figure 11. (**a**) The time series of potential energy anomaly change $\Delta\varphi$ and corresponding time-smoothed result at L1 and (**b**) L2. The shaded bars $d1$, $d2$, and $d3$ represent one-day periods on different dates. (**c**) The distribution of the differences of $\Delta\varphi$ between $d2$ and $d1$, and (**d**) between $d3$ and $d2$.

Figure 11c,d shows the differences in $\Delta\varphi$ between different intervals. For different intervals, these differences are almost in anti-phase, suggesting that the stratification in the Luzon Strait is always being redistributed within a spring-neap cycle. Regarding stratification, this result means that the baroclinic field in the Luzon Strait can be disturbed and redistributed by the interaction between asymmetric barotropic forcing and topography. Due to this redistribution, the energy transfer and the internal wave generation, which strongly depend on stratification, will be affected.

3.3. Energy Transfer

In order to analyze energy transfer and internal wave generation, we introduce the depth-integrated barotropic and baroclinic energy equations [24,41,42].

$$\frac{\partial}{\partial t}(\overline{E_{k0}} + \overline{E_{p0}}) + \nabla \cdot \overline{F_0} = -\overline{C} - \overline{\epsilon_0} \tag{2}$$

$$\frac{\partial}{\partial t}(\overline{E_k'} + \overline{E_p'}) + \nabla \cdot \overline{F'} = \overline{C} - \overline{\epsilon'} \tag{3}$$

where subscript $_0$ and superscript $'$ indicate barotropic and baroclinic, respectively. $\overline{E_k}$ is the kinetic energy, $\overline{E_p}$ is the available potential energy, $\overline{F}$ is the energy flux, $\overline{C}$ is the barotropic to baroclinic energy conversion rate that connects the two equations, and $\overline{\epsilon}$ is the dissipation term including the conversion and radiation processes and bottom drag. As this paper mainly focuses on the baroclinic responses of barotropic forcing, we diagnose only the relative kinetic and potential energy terms and the conversion rate between the two.

$$\overline{C} = p_b' W \tag{4}$$

$$\overline{E_{k0}} = \frac{1}{2}\rho_0 (U^2 + V^2) H \tag{5}$$

$$\overline{E_{p0}} = \frac{1}{2}\rho_0 g \eta^2 \tag{6}$$

$$\overline{E_k'} = \frac{1}{2}\rho_0 \int_{-d}^{\eta} (u'^2 + v'^2 + w^2) dz \tag{7}$$

$$\overline{E_p'} = \frac{g^2}{2\rho_0} \int_{-d}^{\eta} \frac{\rho'^2}{N_b^2} dz \tag{8}$$

where p_b' is the perturbation pressure at the bottom; $W = -\vec{U} \cdot \nabla H$ is the vertical velocity at the bottom due to barotropic flow over variable topography; ρ_0 is the reference density; $u' = u - U$ and $v' = v - V$ are the zonal and meridional baroclinic velocity, respectively; w is the vertical velocity; and $\rho' = \rho - \rho_b$ is the perturbation density due to wave motions, where ρ_b is the background density during the selected spring-neap cycle.

Overall, $\overline{C}$ represents the conversion rate from barotropic to baroclinic mode. Generally, $\overline{C}$ should be the sink term for Equation (2) and the source term for Equation (3), which also means that baroclinic terms gain energy from barotropic terms, on average. However, $\overline{C}$ can be a sink for baroclinic components, for example, $\overline{C}$ becomes negative when W and p_b' are out of phase [24,43]. In our case, $\overline{C}$ is positive most of the time (Figure 12a), which represents energy transfer from barotropic tide to baroclinic tide. Negative $\overline{C}$ can reach nearly half of the maximum positive value during spring tide, which suggests a strong local dissipation of the baroclinic tide for this period.

For kinetic energy, the barotropic part $\overline{E_{k0}}$ and the baroclinic part $\overline{E_k'}$ mainly change according to barotropic and baroclinic velocity, respectively. $\overline{E_{p0}}$ mainly changes with sea-level height and represents the potential energy due to surface waves. Equation (8) is an exact expression for the baroclinic potential energy if the fluid is linearly stratified [44]. In our case, N_b at L1 is slowly varying and almost constant, which suggests that this expression is suitable for evaluating the local available potential energy. $\overline{E_p'}$ directly measures the strength of isopycnal perturbations.

Figure 12b–e shows the changes for those energy components. The time series of $\overline{E_{k0}}$ and $\overline{E_{p0}}$ both show a significant spring-neap cycle and peaks in one day, which matches with the local barotropic tidal signal. The baroclinic energy components exhibit different features. $\overline{E_k'}$ reaches the highest peak when the local barotropic current is directed eastward. However, this behavior changes slowly and does not show a second narrow peak when the local barotropic current is directed westward. Especially for $\overline{E_p'}$, it shows a weaker correlation to the boundary forcing, and thus more nonlinear characteristics comparing to barotropic energy components.

Figure 12f shows the time-smoothed results of each of the above components with the same color. In order to analyze the interaction between the components, we applied a suitable amplification factor. The largest lag difference among energy components is between $\overline{E_p'}$ and the other components. Except for $\overline{E_p'}$, the other three energy components show the same fortnightly variability that matches well with the local barotropic forcing. The lag of $\overline{E_p'}$ indicates that the maximum baroclinic disturbance is not generated during the maximum barotropic forcing. At L1, the lag is approximately two days, which indicates that the maximum baroclinic disturbance occurs two days after the maximum spring tide. As we analyzed before, the baroclinic potential energy and barotropic components exhibit different features. The baroclinic potential energy depends on the perturbation density, which depends on local stratification and dynamic field. The stratification shows a fortnightly variability resulting from the advection of buoyancy by rectified baroclinic flows. In general, the fortnightly stratification variability, i.e., the phase of baroclinic potential energy and the phase of barotropic forcing do not match at most locations. Thus, our result shows a lag or lead relation between the maximum baroclinic potential energy and the maximum barotropic forcing on fortnight time scale. Additionally, the time-smoothed result of $\overline{C}$ shows a comparable lag, which suggests that the maximum barotropic to baroclinic energy conversion is affected by this fortnightly stratification variability.

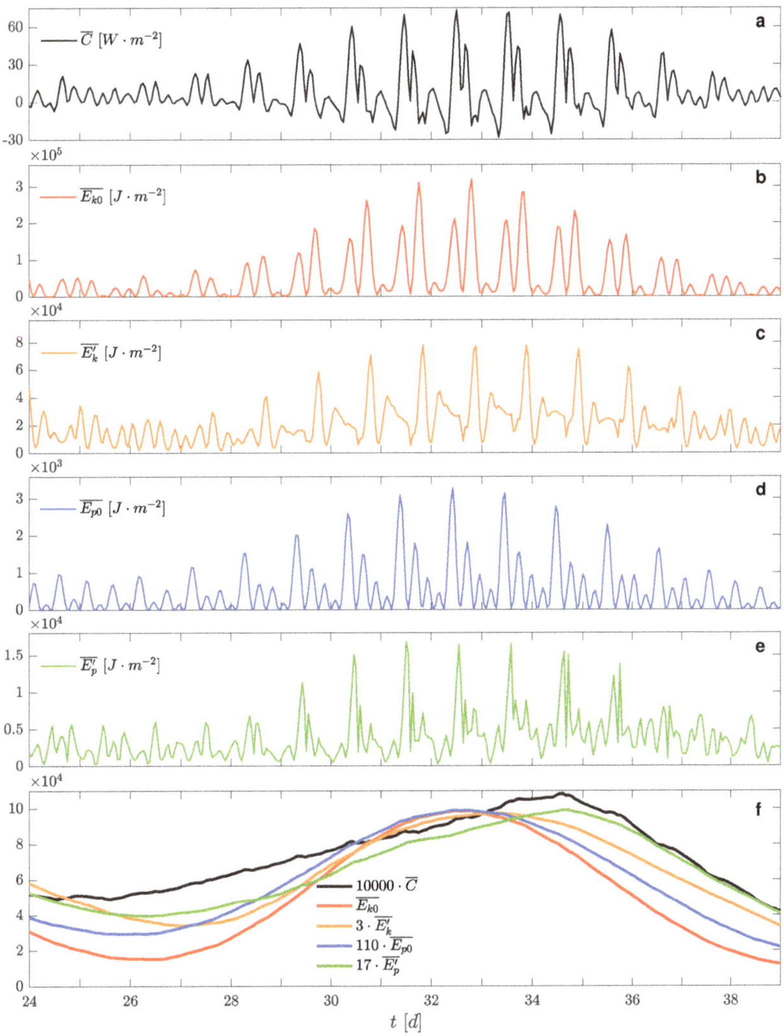

Figure 12. (**a**) Time series of depth-integrated barotropic to baroclinic conversion rate $\overline{C}$, (**b**) barotropic kinetic energy $\overline{E_{k0}}$, (**c**) baroclinic kinetic energy $\overline{E'_k}$, (**d**) barotropic potential energy $\overline{E_{p0}}$, and (**e**) baroclinic potential energy $\overline{E'_p}$ at L1. (**f**) Time-smoothed results of the above-mentioned components with the same color.

Here, we introduce a harmonic fit $f(t) = A\cos(\omega t + P) + M$ of the time-smoothed results, where ω is the selected spring-neap cycle frequency corresponding to 14.78 days, t is the time, A is the amplitude, P is the phase, M is the time average of a spring-neap cycle, and f is the harmonic fitted result. In order to evaluate the goodness of this fit, the coefficient of determination $R = 1 - \sum_{i=1}^{n} \frac{(y_i - f_i)^2}{(y_i - y_{av})^2}$ is calculated here, where f_i is the predicted value from the fit, y_i represents the observed data, and y_{av} is the mean of observed data. R is generally a value between 0 and 1, and a value closer to 1 indicates a better fit.

Figure 13 shows the harmonic fit results of $\overline{E_{k0}}$ and $\overline{C}$ for model days 24 to 38. The average R of $\overline{E_{k0}}$ and $\overline{C}$ in the Luzon Strait is 0.9664 and 0.9097, respectively, which suggests that most of the spring-neap variation of barotropic and baroclinic tides in the Luzon Strait can be explain by this fit. Strong $\overline{E_{k0}}$ is mainly distributed in channels between islands and seamounts (Figure 10c), and is accompanied by intense spring-neap variation (Figure 13a). The main generation sites and dissipation sites of baroclinic tides, suggested by time-averaged M (Figure 13f), are mainly distributed along the two ridges [7,23,31] and are accompanied by intense spring-neap variation as well (Figure 13d). P suggests the arrival time of spring tide. The arrival time of the maximum $\overline{E_{k0}}$ (Figure 13b) and $\overline{C}$ (Figure 13e) during the selected period can be calculated from the phase. For $\overline{E_{k0}}$, the arrival time shows differences along the isobath in general. For $\overline{C}$, only the sites where R is greater than 0.9 and where the average $\overline{C}$ is greater than 2 $W \cdot m^{-2}$ are presented because we mainly focus on the generation process, the arrival time is also quite similar in the main generation sites.

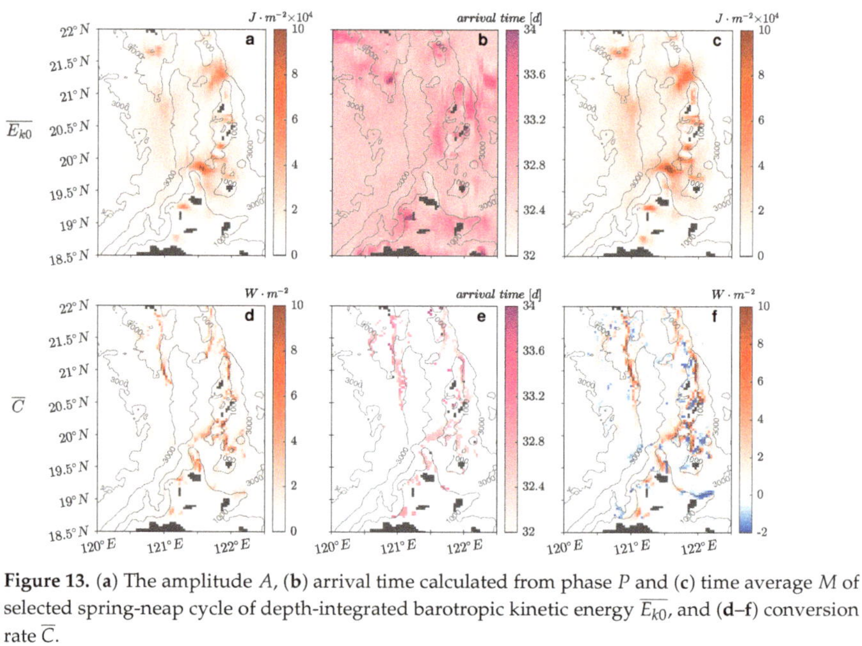

Figure 13. (**a**) The amplitude A, (**b**) arrival time calculated from phase P and (**c**) time average M of selected spring-neap cycle of depth-integrated barotropic kinetic energy $\overline{E_{k0}}$, and (**d–f**) conversion rate $\overline{C}$.

Figure 14 shows the estimated lag between $\overline{C}$ and $\overline{E_{k0}}$ calculated from the phase difference. Only the generation sites are presented as well as Figure 13e. Obviously, on the fortnightly time scale, the $\overline{C}$ is not phase-locked to $\overline{E_{k0}}$ on potential generation sites of the Luzon Strait. Similar to L1, which we analyzed above, there are many other sites where $\overline{C}$ lags behind $\overline{E_{k0}}$. Meanwhile, during this spring-neap period, $\overline{E_{k0}}$ may also lag behind $\overline{C}$ in some locations. The coexistence of both positive and negative lags is an expected result as well as we can see the stratification in the Luzon Strait is fortnightly redistributed (Figure 11). The difference in the arrival time between the maximum barotropic tidal forcing and the maximum barotropic to baroclinic conversion rate are not only a local relation at L1 but are also a more general pattern, which suggests a lead-lag relation between barotropic tidal forcing and maximum barolinic response within the fortnightly tidal cycle.

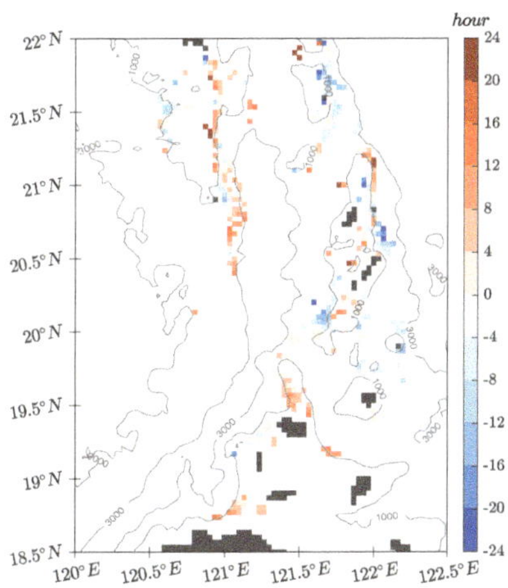

Figure 14. The estimated lag between conversion rate $\overline{C}$ and depth-integrated barotropic kinetic energy $\overline{E_{k0}}$ at main generation sites of baroclinic tides in the Luzon Strait. Positive means $\overline{C}$ lags behind $\overline{E_{k0}}$.

4. Discussion

Internal waves in the South China Sea begin as sinusoidal disturbances in the Luzon Strait [15]. Zhao and Alford presented a simple prediction model that can estimate the arrival times of internal solitary waves in the northeastern South China Sea based on a phase-locked relationship within the tidal cycle between internal solitary waves on the South China Sea shelf and westward tidal currents in the Luzon Strait [9]. However, the amplitude of the internal solitary waves, which depends on the strength of tidal forcing, background flow, and stratification near the generation region, is unpredictable in the prediction model in the above work. There is a clear correlation between the magnitude of semidiurnal tidal current in the Luzon Strait and the corresponding height of the leading solitary wave in the basin of South China Sea [12], which suggests the importance of fortnightly variability in the prediction of internal waves. The intensity of the initial disturbances that can be represented by the baroclinic potential energy at the generation source depends on the local stratification and flow. The local stratification and flow can be modified by Kuroshio and mesoscale eddies in the Luzon Strait, thereby affecting the internal wave generation and evolution [18,45–47]. Except for these two factors, our model results demonstrate that the stratification can also be altered by rectified baroclinic flows, thereby resulting in a lead–lag phenomenon within the fortnightly cycle.

There are some factors that may alter the stratification in our simulation. The two-ridge system in the Luzon Strait enhances the interaction between baroclinic tides from different sites, which may also affect the stratification variation in our model. This effect is considered to be responsible for the high local dissipation [11,24], which should weaken the stratification. Thermal wind balance driven flows caused by the horizontally non-homogeneous stratification may enhance mixing and ultimately weaken the stratification. Mean barotropic flows caused by tidal rectification can also suppress or uplift the stratification and thus modulate the stratification redistribution [36,37]. The strong mean barotropic flows exist only near the ridges, while the redistribution of stratification occurs over the whole Luzon Strait, therefore these mean barotropic flows play a secondary role for modulating the stratification over this research domain.

Although our model grid is too coarse to discuss the details of internal solitary wave generation and our diffusion and viscosity scheme cannot resolve wave-breaking, the energy calculation is still credible as background dissipation is included to parameterize this effect. Meanwhile, the interaction between the asymmetric tidal current and the topography generally exists in the Luzon Strait, and can uplift and suppress the isotherms [18]. Thus, the fortnightly variability of stratification revealed by our model results certainly affects the prediction of internal wave amplitude. Our model results suggest that due to this lead-lag relation, the maximum internal solitary wave under certain conditions may not be triggered by the maximum barotropic forcing.

5. Conclusions

We analyzed the baroclinic responses to barotropic forcing by investigating two aspects, namely, stratification variability and energy transfer, using the MITgcm model [27]. We first validated the model results and thus proved the reliability of our simulation. The stratification variability was investigated by analyzing profiles of temperature and salinity, as well as time series of buoyancy frequency and potential energy anomaly [40]. The energy transfer was investigated based on the depth-integrated barotropic and baroclinic energy equations [24,41,42], specifically by diagnosing the relative kinetic and potential energy terms and the conversion rate in the equations.

We found that the stratification in the Luzon Strait exhibits daily variation caused by daily variations of baroclinic tidal flows and fortnightly variability mainly caused by rectified baroclinic flows. The interaction between asymmetric barotropic forcing and topography generates intense baroclinic flows and thus offers an approach to increasing the stratification without buoyancy inputs like precipitation, fresh water from rivers, and surface heat fluxes. This interaction can also decrease the stratification, thus resulting in the fortnightly variability of stratification. In a scenario without surface buoyancy fluxes, we demonstrated that the stratification in the Luzon Strait can be periodically redistributed by the interaction between periodic asymmetric barotropic forcing and topography. Each barotropic and baroclinic energy component reflects a spring-neap cycle overlaid on the daily variation. The phases of the fortnightly cycle of baroclinic potential energy and conversion rate at L1, which is one location of internal wave generation in the Luzon Strait, do not match the phase of the barotropic energy component, which indicates that the internal wave generation is affected by this fortnightly stratification variability and that the maximum disturbance of these internal waves may not be generated during the maximum barotropic forcing. Extended to the whole Luzon Strait, this lead–lag relation between barotropic tidal forcing and maximum baroclinic response within the fortnightly tidal cycle generally exists in the source of internal waves. In summary, we infer that the fortnightly variability of stratification in the Luzon Strait due to rectified baroclinic flows can significantly affect energy transfer and internal wave generation.

The exact length of the lead–lag relation that determines the accuracy of internal wave amplitude prediction might be affected by the mixing parameterization in our model and by other processes such as mesoscale eddy intrusion, Kuroshio intrusion, and strong upper-layer mixing induced by winds in the real ocean. In order to improve the ability to predict internal waves, each of the effects of the above factors needs further investigation.

Author Contributions: Conceptualization, Z.Z., X.C., and T.P.; methodology, Z.Z.; software, Z.Z.; validation, Z.Z., X.C., and T.P.; formal analysis, Z.Z.; investigation, Z.Z.; resources, X.C and T.P.; data curation, Z.Z.; writing—original draft preparation, Z.Z.; writing—review and editing, Z.Z., X.C., and T.P.; visualization, Z.Z.; supervision, X.C. and T.P.; project administration, X.C.; funding acquisition, X.C. and T.P. All authors have read and agreed to the published version of the manuscript.

Funding: This research was supported by National Natural Science Foundation of China (no. U1706218), and the German BMBF CLISTORM under grant no. 03F0781A.

Data Availability Statement: The model results presented in this study are available on request from the corresponding author.

Acknowledgments: We acknowledge the National Supercomputing Center of Jinan for providing the computing resources. The bathymetric data were obtained from the general bathymetric chart of the oceans 2008 http://www.gebco.net (last access: 26 June 2021). The barotropic forcing data were obtained from the OSU TOPEX/Poseidon Global Inverse Solution 7.2 http://g.hyyb.org/archive/Tide/TPXO/TPXO_WEB/global.html (last access: 26 June 2021). The initial field data were obtained from World Ocean Atlas 2009 https://www.nodc.noaa.gov/OC5/WOA09/pr_woa09.html (last access: 26 June 2021). Two anonymous reviewers provided numerous helpful suggestions for improving the manuscript.

Conflicts of Interest: The authors declare no conflict of interest.

References

1. Egbert, G.D.; Ray, R.D. Significant dissipation of tidal energy in the deep ocean inferred from satellite altimeter data. *Nature* **2000**, *405*, 775–778. [CrossRef]
2. Laurent, L.S.; Garrett, C. The role of internal tides in mixing the deep ocean. *J. Phys. Oceanogr.* **2002**, *32*, 2882–2899. [CrossRef]
3. Alford, M.H. Redistribution of energy available for ocean mixing by long-range propagation of internal waves. *Nature* **2003**, *423*, 159–162. [CrossRef]
4. Buijsman, M.C.; Stephenson, G.; Ansong, J.; Arbic, B.; Green, J.; Richman, J.; Shriver, J.; Vic, C.; Wallcraft, A.; Zhao, Z. On the interplay between horizontal resolution and wave drag and their effect on tidal baroclinic mode waves in realistic global ocean simulations. *Ocean. Model.* **2020**, *152*, 101656. [CrossRef]
5. Klymak, J.M.; Pinkel, R.; Liu, C.-T.; Liu, A.K.; David, L. Prototypical solitons in the south china sea. *Geophys. Res. Lett.* **2006**, *33*, l11607. [CrossRef]
6. Huang, X.; Chen, Z.; Zhao, W.; Zhang, Z.; Zhou, C.; Yang, Q.; Tian, J. An extreme internal solitary wave event observed in the northern south china sea. *Sci. Rep.* **2016**, *6*, 30041. [CrossRef]
7. Niwa, Y.; Hibiya, T. Three-dimensional numerical simulation of m^2 internal tides in the east china sea. *J. Geophys. Res. Ocean.* **2004**, *109*, c04027. [CrossRef]
8. Zhao, Z.; Klemas, V.; Zheng, Q.; Yan, X.-H. Remote sensing evidence for baroclinic tide origin of internal solitary waves in the northeastern south china sea. *Geophys. Res. Lett.* **2004**, *31*, l06302. [CrossRef]
9. Zhao, Z.; Alford, M.H. Source and propagation of internal solitary waves in the northeastern south china sea. *J. Geophys. Res. Ocean.* **2006**, *111*, c11012. [CrossRef]
10. Jan, S.; Chern, C.-S.; Wang, J.; Chao, S.-Y. Generation of diurnal k1 internal tide in the luzon strait and its influence on surface tide in the south china sea. *J. Geophys. Res. Ocean.* 2007, *112*, c06019. [CrossRef]
11. Alford, M.H.; MacKinnon, J.A.; Nash, J.D.; Simmons, H.; Pickering, A.; Klymak, J.M.; Pinkel, R.; Sun, O.; Rainville, L.; Musgrave, R.; et al. Energy flux and dissipation in luzon strait: Two tales of two ridges. *J. Phys. Oceanogr.* **2011**, *41*, 2211–2222. [CrossRef]
12. Li, Q.; Farmer, D.M. The generation and evolution of nonlinear internal waves in the deep basin of the south china sea. *J. Phys. Oceanogr.* **2011**, *41*, 1345–1363. [CrossRef]
13. Zhang, Z.; Fringer, O.B.; Ramp, S.R. Three-dimensional, nonhydrostatic numerical simulation of nonlinear internal wave generation and propagation in the south china sea. *J. Geophys. Res. Ocean.* **2011**, *116*, c05022. [CrossRef]
14. Guo, C.; Vlasenko, V.; Alpers, W.; Stashchuk, N.; Chen, X. Evidence of short internal waves trailing strong internal solitary waves in the northern south china sea from synthetic aperture radar observations. *Remote. Environ.* **2012**, *124*, 542–550. [CrossRef]
15. Alford, M.H.; Peacock, T.; MacKinnon, J.A.; Nash, J.D.; Buijsman, M.C.; Centuroni, L.R.; Chao, S.-Y.; Chang, M.-H.; Farmer, D.M.; Fringer, O.B.; et al. The formation and fate of internal waves in the south china sea. *Nature* **2015**, *521*, 65–69. [CrossRef]
16. Ramp, S.R.; Yang, Y.J.; Bahr, F.L. Characterizing the nonlinear internal wave climate in the northeastern south china sea. *Nonlinear Processes Geophys.* **2010**, *17*, 481–498. [CrossRef]
17. Xu, Z.; Yin, B.; Hou, Y.; Liu, A.K. Seasonal variability and north–south asymmetry of internal tides in the deep basin west of the luzon strait. *J. Mar. Syst.* **2014**, *134*, 101–112. [CrossRef]
18. Buijsman, M.C.; Kanarska, Y.; McWilliams, J.C. On the generation and evolution of nonlinear internal waves in the south china sea. *J. Geophys. Res. Ocean.* 2010, *115*, C02012. [CrossRef]
19. Lee, C.-Y.; Beardsley, R.C. The generation of long nonlinear internal waves in a weakly stratified shear flow. *J. Geophys. Res.* **1974**, *79*, 453–462. [CrossRef]
20. Maxworthy, T. A note on the internal solitary waves produced by tidal flow over a three-dimensional ridge. *J. Geophys. Res. Ocean.* **1979**, *84*, 338–346. [CrossRef]
21. Maxworthy, T. On the formation of nonlinear internal waves from the gravitational collapse of mixed regions in two and three dimensions. *J. Fluid Mech.* **1980**, *96*, 47–64. [CrossRef]
22. Gerkema, T. Internal and interfacial tides: Beam scattering and local generation of solitary waves. *J. Mar. Res.* **2001**, *59*, 227–255. [CrossRef]
23. Jan, S.; Lien, R.-C.; Ting, C.-H. Numerical study of baroclinic tides in luzon strait. *J. Oceanogr.* **2008**, *64*, 789–802. [CrossRef]
24. Buijsman, M.C.; Legg, S.; Klymak, J. Double-ridge internal tide interference and its effect on dissipation in luzon strait. *J. Phys. Ocean.* **2012**, *42*, 1337–1356. [CrossRef]

25. Baines, P. On internal tide generation models. *Deep. Sea Res. Part Oceanogr. Res. Pap.* **1982**, *29*, 307–338. [CrossRef]
26. Xing, J.; Davies, A.M. Influence of stratification and topography upon internal wave spectra in the region of sills. *Geophys. Res. Lett.* **2006**, *33*, L23606. [CrossRef]
27. Marshall, J.; Adcroft, A.; Hill, C.; Perelman, L.; Heisey, C. A finite-volume, incompressible navier stokes model for studies of the ocean on parallel computers. *J. Geophys. Res. Ocean.* **1997**, *102*, 5753–5766. [CrossRef]
28. Large, W.G.; McWilliams, J.C.; Doney, S.C. Oceanic vertical mixing: A review and a model with a nonlocal boundary layer parameterization. *Rev. Geophys.* **1994**, *32*, 363–403. [CrossRef]
29. Fang, G. Tide and tidal current charts for the marginal seas adjacent to china. *Chin. J. Oceanol. Limnol.* **1986**, *4*, 1–16.
30. Zu, T.; Gan, J.; Erofeeva, S.Y. Numerical study of the tide and tidal dynamics in the south china sea. *Deep. Sea Res. Part Oceanogr. Res. Pap.* **2008**, *55*, 137–154. [CrossRef]
31. Li, M.; Hou, Y.; Li, Y.; Hu, P. Energetics and temporal variability of internal tides in luzon strait: A nonhydrostatic numerical simulation. *Chin. J. Oceanol. Limnol.* **2012**, *30*, 852–867. [CrossRef]
32. Ray, R.D.; Mitchum, G.T. Surface manifestation of internal tides generated near hawaii. *Geophys. Res. Lett.* **1996**, *23*, 2101–2104. [CrossRef]
33. Cummins, P.F.; Oey, L.-Y. Simulation of barotropic and baroclinic tides off northern british columbia. *J. Phys.* **1997**, *27*, 762–781. [CrossRef]
34. Boon, J.D.; Byrne, R. On basin hypsometry and the morphodynamic response of coastal inlet systems. *Mar. Geol.* **1981**, *40*, 27–48. [CrossRef]
35. Friedrichs, C.T.; Aubrey, D. Non-linear tidal distortion in shallow well-mixed estuaries: a synthesis. *Estuar. Coast. Shelf Sci.* **1988**, *27*, 521–545. [CrossRef]
36. Loder, J.W. Topographic rectification of tidal currents on the sides of Georges Bank. *J. Phys. Oceanogr.* **1980**, *10*, 1399–1416. [CrossRef]
37. Chen, C.S; Lai, Z.; Beardsley, R.; Xu, Q.; Lin, H.; Viet, N. Current separation and upwelling over the southeast shelf of Vietnam in the South China Sea. *J. Geophys. Res. Ocean.* **2012**, *117*, C03033. [CrossRef]
38. Pickering A.; Alford, M.; Nash, J.; Rainville, L.; Ko, D.S.; Lim, B. Structure and Variability of Internal Tides in Luzon Strait. *J. Phys. Oceanogr.* **2015**, *45*, 1574–1594. [CrossRef]
39. Simpson, J.H.; Crisp, D.J.; Hearn, C. The shelf-sea fronts: Implications of their existence and behaviour [and discussion]. *Philos. Trans. R. Soc. Lond. Ser. A Math. Phys. Sci.* **1981**, *302*, 531–546.
40. de Boer, G.J.; Pietrzak, J.D.; Winterwerp, J.C. Using the potential energy anomaly equation to investigate tidal straining and advection of stratification in a region of freshwater influence. *Ocean. Model.* **2008**, *22*, 1–11. [CrossRef]
41. Kurapov, A.L.; Egbert, G.D.; Allen, J.S.; Miller, R.N.; Erofeeva, S.Y.; Kosro, P.M. The m2 internal tide off oregon: Inferences from data assimilation. *J. Phys. Oceanogr.* **2003**, *33*, 1733–1757. [CrossRef]
42. Kang, D.; Fringer, O. Energetics of barotropic and baroclinic tides in the monterey bay area. *J. Phys. Oceanogr.* **2012**, *42*, 272–290. [CrossRef]
43. Kelly, S. M.; Nash, J.D. Internal-tide generation and destruction by shoaling internal tides. *Geophys. Res. Lett.* **2010**, *37*, l23611. [CrossRef]
44. Kang, D.; Fringer, O. On the calculation of available potential energy in internal wave fields. *J. Phys. Oceanogr.* **2010**, *40*, 2539–2545. [CrossRef]
45. Jan, S.; Chern, C.-S.; Wang, J.; Chiou, M.-D. Generation and propagation of baroclinic tides modified by the Kuroshio in the Luzon Strait. *J. Geophys. Res. Ocean.* **2012**, *117*, c02019. [CrossRef]
46. Ma, Barry B.; Lien, R.C.; Ko, D.S. The variability of internal tides in the Northern South China Sea. *J. Oceanogr.* **2013**, *59*, 619–630. [CrossRef]
47. Li, Q.; Wang, B.; Chen, X.; Chen, X.E.; Park, J.H. Variability of nonlinear internal waves in the South China Sea affected by the Kuroshio and mesoscale eddies. *J. Geophys. Res. Ocean.* **2016**, *121*, 2098–2118. [CrossRef]

Journal of
Marine Science and Engineering

Article

Near-Inertial Waves Induced by Typhoon Megi (2010) in the South China Sea

Anzhou Cao [1,2,3,*], Zheng Guo [4], Yunhe Pan [1], Jinbao Song [1], Hailun He [2] and Peiliang Li [1,5]

1 Ocean College, Zhejiang University, Zhoushan 316021, China; 18716037312@163.com (Y.P.); songjb@zju.edu.cn (J.S.); lipeiliang@zju.edu.cn (P.L.)
2 State Key Laboratory of Satellite Ocean Environment Dynamics, Second Institute of Oceanography, Ministry of Natural Resources, Hangzhou 310012, China; hehailun@sio.org.cn
3 Laboratory for Regional Oceanography and Numerical Modeling, Qingdao National Laboratory for Marine Science and Technology, Qingdao 266061, China
4 Marine Science and Technology College, Zhejiang Ocean University, Zhoushan 316022, China; guozheng-gz@163.com
5 Hainan Institution of Zhejiang University, Sanya 572025, China
* Correspondence: caoanzhou@zju.edu.cn

Abstract: Near-inertial waves (NIWs) are a kind of internal wave, which are usually generated by synoptic wind forcing and play an important role in the oceanic energy budget. However, the lack of in situ observations limits our understanding of NIWs to some extent. Through a comparison with in situ observations, in this study, we first showed that the hybrid coordinate ocean model reanalysis results could reasonably reproduce the typhoon-induced NIWs, and we then adopted these data to investigate the NIWs induced by typhoon Megi in 2010 in the South China Sea (SCS). The results indicate that Megi-induced near-inertial kinetic energy was mainly concentrated in the SCS Basin. In the vertical direction, Megi-induced NIWs could propagate to 1000 m depth. The damping and modal content of Megi-induced NIWs were site-dependent: In the region near Megi's track, NIWs were dominated by the first three baroclinic modes and damped quickly; whereas in two zones to the west of the Luzon Island and Luzon Strait, the e-folding time of Megi-induced NIWs could be longer than 20 days and higher modes (mode-4 to mode-7) were enhanced several days after the passage of Megi. Possible mechanisms of these phenomena were also explored in this study.

Keywords: near-inertial waves; typhoon Megi; South China Sea; hybrid coordinate ocean model reanalysis results

Citation: Cao, A.; Guo, Z.; Pan, Y.; Song, J.; He, H.; Li, P. Near-Inertial Waves Induced by Typhoon Megi (2010) in the South China Sea. *J. Mar. Sci. Eng.* **2021**, *9*, 440. https://doi.org/10.3390/jmse9040440

Academic Editors: Matt Lewis and SungHyun Nam

Received: 8 March 2021
Accepted: 15 April 2021
Published: 18 April 2021

1. Introduction

Near-inertial waves (NIWs) are a kind of internal wave, which are ubiquitous in the global ocean. In the internal wave spectrum, NIWs appear as a predominant peak near the local inertial frequency, and hence, act as a dominant mode of high-frequency variability in the ocean [1]. NIWs play an important role in the oceanic energy budget [2,3], as they can cause intense shear [4,5] and contribute to the elevation of turbulent mixing [6,7].

There are various mechanisms that can cause NIWs, among which, the most important is synoptic wind forcing. According to previous estimations, the global power of wind-driven near-inertial motions is 0.3–1.5 TW [8–14], which is comparable to the global power converted from astronomical tides to internal tides [3,15,16]. Due to strong wind stresses and their compact size, tropical cyclones (typhoons and hurricanes) are an efficient generator of NIWs [1]. Moreover, nonlinear wave–wave interaction, including parametric subharmonic instability [17–19] and resonant triad interaction [20–22], lee waves [23], frontal jets [24] and mesoscale eddies [25], can also induce NIWs under some conditions.

The Western North Pacific is the region with the highest concentration of typhoons in the world [26]. According to the estimation of Nguyen et al. [27], there are an average of 22 typhoons per year and most of them pass through the South China Sea (SCS), the largest marginal sea in the Western North Pacific. The frequency of typhoons has shown an increasing trend in the SCS [28]. They import a significant amount of energy to the ocean and induce intense NIWs [29]. Several previous studies have reported typhoon-induced NIWs in the SCS through analyzing in situ observations [30–40]. However, due to the differences in typhoon characteristics, mooring-measuring ranges, distances between typhoon centers and moorings, and local conditions, the NIWs induced by different typhoons usually exhibit different features and those induced by the same typhoon are site-dependent [31,32,34,35,37]. More importantly, because of the great difficulties and large costs of in situ observations, it is nearly impossible to simultaneously deploy sufficient moorings to observe NIWs induced by the same typhoon, which limits our understanding of NIWs to some extent.

In this study, through a comparison with in situ observations, we first showed that the hybrid coordinate ocean model (HYCOM) reanalysis results could reasonably reproduce the typhoon-induced NIWs. Thereafter, these data were used to investigate the characteristics of NIWs induced by typhoon Megi in 2010, with the aim of deepening our understanding of typhoon-induced NIWs in the SCS. The paper is organized as follows. Typhoon Megi, the HYCOM reanalysis results and corresponding data analysis methods are introduced in Section 2. A comparison between the HYCOM reanalysis results and in situ observations is performed in Section 3. After validation, the HYCOM reanalysis results are used to reveal the energy and modal characteristics of Megi-induced NIWs, which are shown in Sections 4 and 5, respectively. A discussion is presented in Section 6. Finally, the conclusions of this study complete the paper in Section 7.

2. Data and Methodology

2.1. Typhoon Megi

Megi was the 15th typhoon that occurred in 2010 in the Western Pacific. It was also the strongest typhoon worldwide in 2010 [32]. According to the best track data from the China Meteorological Administration tropical cyclone database [41], Megi first formed as a tropical depression at 11.8° N, 141.4° E on 13 October. Then, it moved northwestward and enhanced rapidly. Before landing on Luzon Island on 18 October, Megi upgraded to a super typhoon. After passing Luzon Island, Megi weakened a little to a severe typhoon until 22 October. Thereafter, Megi damped quickly and finally dissipated on 24 October. The track and status of Megi in the SCS are shown in Figure 1.

2.2. Data

The HYCOM reanalysis results (GLBu0.08/expt_19.1, including surface elevation, horizontal velocities, water temperature and salinity) in the SCS from 16 October to 20 November 2010 were used to analyze the ocean dynamical response to typhoon Megi. These data have a spatial resolution of 1/12.5° and a temporal interval of 3 h, which are available from www.hycom.org/data/glbu0pt08/expt-19pt1 (accessed on 1 January 2021). The wind forcing used in the HYCOM is the 1-hourly National Centers for Environmental Prediction (NCEP) Climate Forecast System Reanalysis (CFSR) wind data, which captured typhoon Megi [37]. Additionally, note that the Navy Coupled Ocean Data Assimilation (NCODA) system was used for data assimilation when generating the reanalysis data. In addition, to validate the HYCOM reanalysis results, the bandpass filtered NIWs at mooring UIB6 (Figure 1) were used in this study, which are cited from [37].

2.3. Methodology

First of all, the HYCOM reanalysis results were compared with in situ observations at mooring UIB6 for validation. A continuous wavelet transform was first performed to demonstrate the existence of NIWs. As tidal forcing was not considered in the HYCOM reanalysis results (GLBu0.08/expt_19.1), the bandpass filtered NIWs rather than the raw currents were compared with those from in situ observations. To be consistent with previous studies [32,37], the fourth-order Butterworth filter was adopted and the cutoff frequency of bandpass filtering was set to [0.58, 0.81] cpd corresponding to 0.80–1.13 times the local Coriolis frequency.

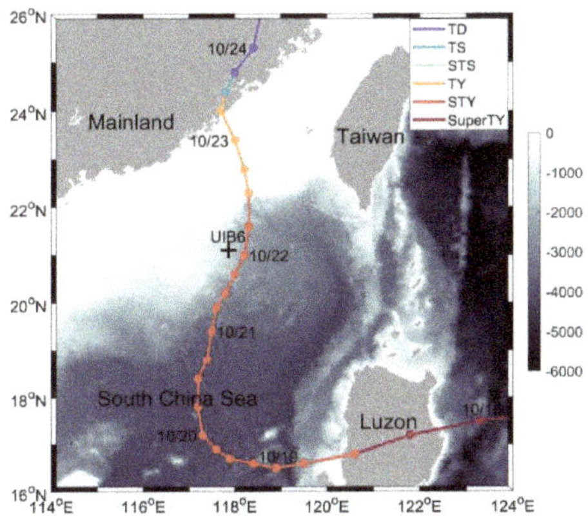

Figure 1. Bathymetry (shading, unit: m) of the northern SCS, track of typhoon Megi (colored lines) and position of mooring UIB6 (black plus). Six-hour positions of Megi's center are denoted by dots and the time at 00:00 from 19 to 24 October is labeled. Intensity of Megi is represented by various colors, as shown in the legend. TD, TS, STS, TY, STY and SuperTY are abbreviations of tropical depression (10.8–17.1 m/s), tropical storm (17.2–24.4 m/s), severe tropical storm (24.5–32.6 m/s), typhoon (32.7–41.4 m/s), severe typhoon (41.5–50.9 m/s) and super typhoon (>51.0 m/s), respectively.

After validation, the near-inertial kinetic energy density (NIKE) was calculated as:

$$\text{NIKE} = \frac{1}{2}\rho_0\left(u_f^2 + v_f^2\right) \tag{1}$$

where ρ_0 = 1024 kg/m^3 is the reference density, u_f and v_f are the zonal and meridional components of the bandpass filtered NIWs, respectively [32]. In this study, we also calculated the depth-integrated NIKE to investigate the horizontal distribution of Megi-induced NIWs. Based on the depth-integrated NIKE, the e-folding time of Megi-induced NIWs was calculated to explore their decay.

To investigate the propagation of Megi-induced NIWs, we adopted the same method as [42,43] to separate the NIWs propagating in different directions. This method is based on Hilbert transform as well as filtering, Fourier transform and its inverse transform. For the time series of one-dimensional, two-dimensional and three-dimensional wave fields, this method can automatically identify waves propagating in two, four and eight directions, respectively. Refer to [42] for details of this method. In this study, we adopted this method to deal with the bandpass filtered NIWs along 118° E to explore the propagation of Megi-induced NIWs and their reflection at the continental slope of the northern SCS.

Modal content is an important characteristic of internal waves [44–48]. Therefore, it was investigated for Megi-induced NIWs in this study. For the zonal and meridional components of NIWs,

$$\begin{cases} u_f(z,t) = \sum\limits_{n=0}^{N_m} u_{fn}(t) \cdot \Pi_n(z) \\ v_f(z,t) = \sum\limits_{n=0}^{N_m} v_{fn}(t) \cdot \Pi_n(z) \end{cases} \tag{2}$$

where u_{fn} and v_{fn} are the modal components of u_f and v_f with respect to mode n (n = 0, 1, ..., N_m, n = 0 for the barotropic mode and n > 0 for baroclinic modes), and

$$\Pi_n(z) = \rho_0 c_n^2 \frac{\mathrm{d}\Phi_n(z)}{\mathrm{d}z} \tag{3}$$

are the normal modes corresponding to velocity, where $\Phi_n(z)$ are the eigenfunctions of the eigenvalue problem for eigenspeed c_n:

$$\frac{d^2\Phi_n}{dz^2} + \frac{N^2}{c_n^2}\Phi_n = 0 \tag{4}$$

subject to boundary conditions $\Phi_n(0) = \Phi_n(-H) = 0$, where H is the water depth and N is the buoyancy frequency [45].

Based on the temperature and salinity data of the HYCOM reanalysis results, the buoyancy frequency was calculated, and hence, normal modes Φ_n and Π_n. According to [47], the time-varying stratification has little influence on the modal decomposition result. Therefore, the time-averaged temperature and salinity were used to calculate the buoyancy frequency and normal modes Φ_n and Π_n. Figure 2 shows an example at 117.04° E, 18.48° N. In theory, N_m should be infinite in modal decomposition. Whereas in practice, N_m is usually set to be a certain value that is sufficient to capture the internal wave features. Generally, according to previous studies, 3, 5 and 10 are typical values used in modal decomposition [45–48]. Additionally, note that too large N_m may cause overfitting [46]. In this study, we found that N_m = 10 could well reproduce the NIKE and did not cause overfitting. Therefore, N_m = 10 was adopted in the modal decomposition in this study.

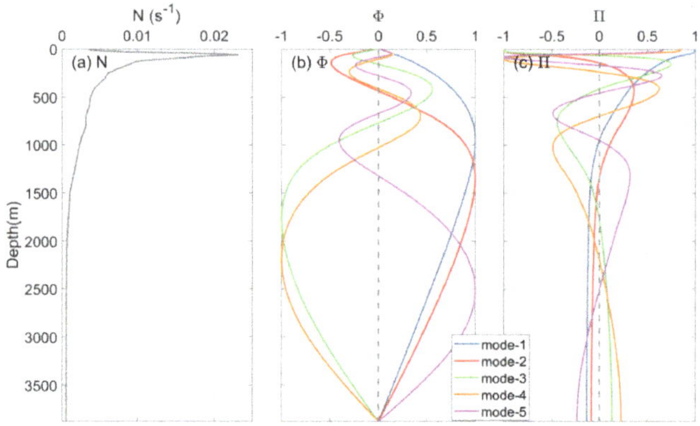

Figure 2. (**a**) Buoyancy frequency and normal modes (**b**) Φ and (**c**) Π of the first five baroclinic modes at 117.04° E, 18.48° N. Note that the normal modes shown in (**b**,**c**) have been normalized.

3. Comparison with In Situ Observations

Figure 3 displays the HYCOM zonal currents at mooring UIB6 as well as the continuous wavelet transformation of HYCOM zonal currents at 400 m depth at the mooring. It is clearly shown, that after the passage of typhoon Megi, oscillations appeared in the zonal currents. In Figure 3b, a peak exists near the local Coriolis frequency, suggesting that the oscillations in the zonal currents are Megi-induced NIWs.

Figure 4 compares the NIWs extracted from in situ observations and the HYCOM reanalysis results at mooring UIB6. Although the temporal intervals of the HYCOM reanalysis results (3 h) and observations (1 h) are different, the NIWs extracted from the HYCOM reanalysis results show a good agreement with those from observations: Both NIWs were rapidly enhanced after the passage of typhoon Megi, suggesting that they were induced by typhoon Megi; both NIWs had upward-propagating phases, suggesting that their energy was downward-propagating, which is consistent with the general features of typhoon-induced NIWs [49–51]; both NIWs were quickly damped at mooring UIB6. This result preliminarily verifies the accuracy of the HYCOM reanalysis results. To quantitatively assess the HYCOM reanalysis results, Figure 5 shows the lowpass filtered current variance of NIWs (Var $= u_f^2 + v_f^2$; [37]) averaged between 50 and 420 m depth (the effective measuring range at UIB6; [37]) for both observations and HYCOM reanalysis results. As seen, both the HYCOM and observed current variance shows the development and decaying of Megi-induced NIWs at UIB6. The correlation coefficient between the HYCOM and observed current variance is 0.97 with a p-value much smaller than 0.01, suggesting good consistency between them. Additionally, note that the peak value of the HYCOM current variance is greater than that of the observed current variance. This is reasonable because tidal forcing was not considered in the HYCOM reanalysis results (GLBu0.08/expt_19.1); hence, nonlinear interaction between NIWs and internal tides cannot occur. As the nonlinear interaction between internal waves can transfer a significant amount of energy [52], the ignored tidal forcing in the HYCOM reanalysis results (GLBu0.08/expt_19.1) finally leads to an overestimation of the energy and current variance of NIWs. Although the HYCOM reanalysis results overestimate the intensity of NIWs, it may only have a limited influence on the distribution, decaying and modal content of NIWs, which are the main focus of this study.

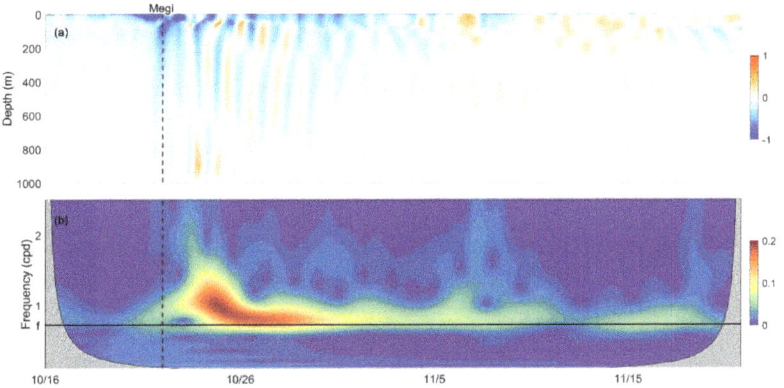

Figure 3. (**a**) HYCOM zonal currents at UIB6 (shading, unit: m/s) as a function of time and depth. (**b**) Continuous wavelet transform (shading, unit: m/s) of HYCOM zonal currents at 400 m depth at UIB6 as a function of time and frequency. The horizontal black solid line indicates the local inertial frequency. In (**a**,**b**), the vertical black dashed lines denote the time when Megi passed (00:00 on 22 October).

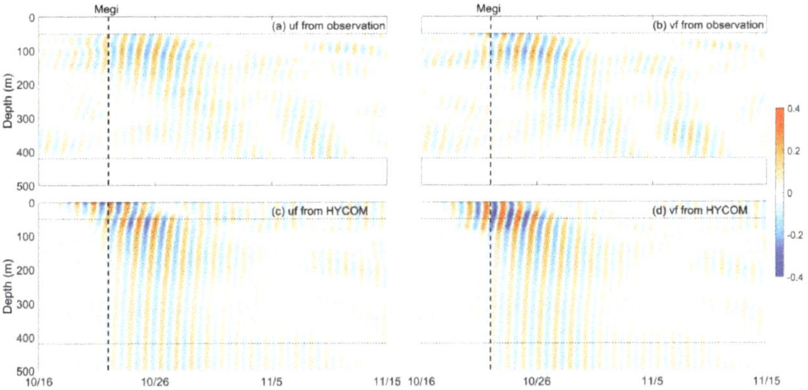

Figure 4. Comparison of (**a**,**c**) zonal and (**b**,**d**) meridional currents of NIWs (shading, unit: m/s) between (**a**,**b**) observations and (**c**,**d**) HYCOM reanalysis results at UIB6.The vertical black dashed line in each subfigure denotes the time when Megi passed (00:00 on 22 October).

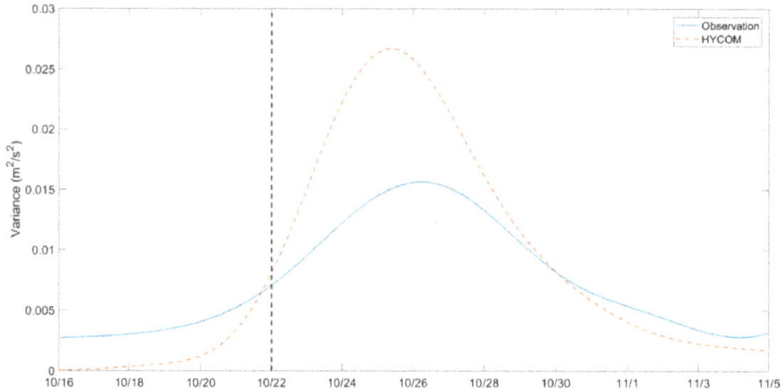

Figure 5. Comparison of lowpass filtered depth-averaged current variance of NIWs between observations (blue solid) and HYCOM reanalysis results (orange dashed) at UIB6. The vertical black dashed line indicates the time when Megi passed (00:00 on 22 October).

4. Megi-Induced NIKE

4.1. Temporal Variation and Spatial Distribution

Figure 6 illustrates the depth-integrated NIKE from 18 October to 1 November with an interval of 2 days, from which the evolution of Megi-induced NIWs is detected. At 00:00 on 18 October when Megi's center did not enter the SCS (Figure 1), slight NIKE was found to the west of Luzon Island (Figure 6a). At 00:00 on 20 October, when Megi's center was at 117.3° E, 17.2° N in the SCS, Megi-induced NIKE reached 10–20 kJ/m^2 and was mainly concentrated in two zones to the north and south of typhoon Megi, respectively (Figure 6b). From 22 to 26 October, strong NIKE with several hot spots exceeding 35 kJ/m^2 appeared along Megi's wake (Figure 6c–e). It should be noted that Megi had left the SCS Basin before 12:00 on 22 October and finally dissipated on 24 October, which means that the strongest NIKE appeared several days after the passage of Megi rather than under its influence. Thereafter, strong NIKE scattered (mainly westward propagated) and the NIKE along Megi's wake quickly damped to below 10 kJ/m^2 after 30 October (Figure 6f–h). According to [53], the ocean's response to a typhoon in the northern hemisphere usually exhibits rightward biased features, i.e., larger sea surface temperature cooling, greater currents and deeper mixed layer appear to the right of the typhoon track. As shown in Figure 6, Megi-induced NIWs also exhibited apparent rightward biased features: the NIKE to the right of Megi's wake was stronger than that to the left.

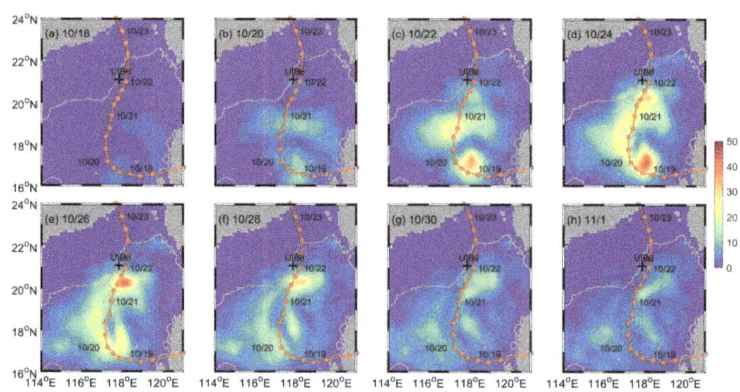

Figure 6. Depth-integrated NIKE (shading, unit: kJ/m^2) at 00:00 on (**a**) 18, (**b**) 20, (**c**) 22, (**d**) 24, (**e**) 26, (**f**) 28 and (**g**) 30 October and (**h**) 1 November 2010. Orange curves denote the track of typhoon Megi and the black plus indicates the position of mooring UIB6. The gray curve represents the 1000 m isobath.

It is interesting to find from Figure 6 that Megi-induced NIKE was mainly concentrated in the deep SCS Basin where the water depth is greater than 1000 m (the gray curve in Figure 6), although typhoon Megi passed over both the deep SCS basin and shallow continental shelf and slope in the northern SCS. To investigate the possible cause of this phenomenon, Figure 7 illustrates several snapshots of meridional currents of NIWs along 118° E on 24 October. Similar results can be found at other meridional sections on the adjacent several days. From Figure 7, we can detect that the continental slope in the northern SCS is supercritical to Megi-induced NIWs, i.e., the topographic slope (the black solid lines in Figure 7) is apparently larger than that of NIWs. In other words, the NIWs impinging on the continental slope in the northern SCS would be reflected to the SCS Basin. However, due to the complex vertical pattern of NIWs shown in Figure 7, the reflection of NIWs is not visible. Therefore, we adopted the same method as [42,43] to separate the NIWs propagating in different directions. Figure 8 shows an example at 00:00 on 24 October. The NIW component with $m > 0$ (m represents the vertical wavenumber) dominates over that with $m < 0$, suggesting that the energy of Megi-induced NIWs mainly propagated downward, which is consistent with the above analysis and general features of typhoon-induced NIWs [49–51]. For the same m, the NIW component with $l > 0$ (l represents the horizontal wavenumber) is comparable to that with $l < 0$, suggesting that the northward-propagating and southward-propagating NIWs had comparable intensity. In order to study whether Megi-induced NIWs reflected on the continental slope, attention should be paid to the northward–downward (l>0 and $m > 0$) and southward–downward ($l < 0$ and $m > 0$) components (Figure 8b,d). According to Figure 8b, two northward–downward-propagating beams impinged on the continental slope in the northern SCS, which radiated from the surface at 19–20.5° N (positive sign) and 20.5–22° N (negative sign), respectively. At the same time, two southward–downward-propagating beams were found to radiate from the continental slope with different signs (Figure 8d). To demonstrate that the two southward–downward-propagating beams are the reflected beams of the two northward–downward-propagating ones, a simple ray tracing model [54,55] is adopted to qualitatively illustrate the propagating paths of NIWs. Given that reflection occurred in a small region near the continental slope in the northern SCS, the beta effect of Coriolis frequency and the influence of background currents were not taken into consideration. The ray tracing model [54,55] is described as

$$\frac{dz}{dy} = \tan\alpha = \sqrt{\frac{\omega^2 - f^2}{N(z)^2 - \omega^2}} \tag{5}$$

where y and z are the local Cartesian coordinates (positive northward and upward, respectively), α is the slope of NIWs, ω, f and $N(z)$ are the NIW, local Coriolis and buoyancy frequencies, respectively. In this study, $N(z)$ was set as the averaged buoyancy frequency near the continental slope along 118° E, f was set to the value of the generation point at the surface and ω was determined through a series of trials. It is found that $\omega = 1.015f$ could lead to reasonable results, therefore, $\omega = 1.015f$ was adopted in the ray tracing model in this study. Figure 8b,d shows the ray tracing results. The two northward–downward-propagating and two southward–downward-propagating beams mentioned above show good consistency with the ray tracing results, confirming the occurrence of reflection of NIWs at the supercritical continental slope in the northern SCS.

Based on the aforementioned analysis, it can be concluded that because the continental slope in the northern SCS is supercritical to Megi-induced NIWs, the NIWs impinging on the continental slope were reflected back and then trapped in the SCS Basin, which finally resulted in the concentrated NIKE in the deep SCS Basin (Figure 6). In addition, because mooring UIB6 was located on the continental slope in the northern SCS, the NIKE here was not very significant. This explains why the strongest typhoon in 2010, Megi, did not generate stronger NIWs than another typhoon, Meranti, at mooring UIB6 [37].

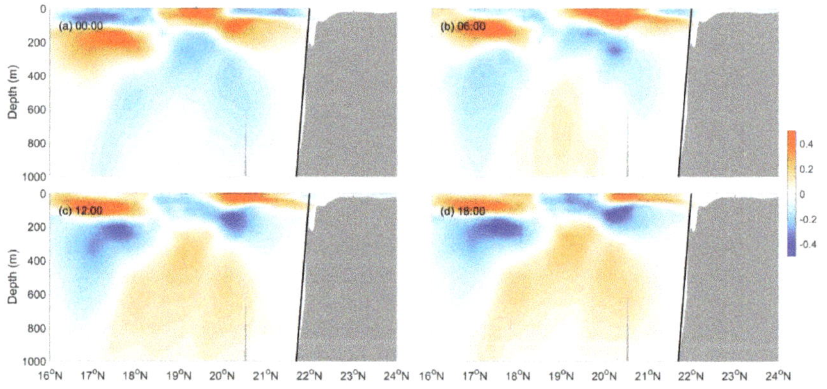

Figure 7. Meridional currents of NIWs (shading, unit: m/s) along 118° E at(**a**) 00:00, (**b**) 06:00, (**c**) 12:00 and (**d**) 18:00 on 24 October. The gray shading in each subfigure indicates the topography. The black solid line in each subfigure indicates the approximate topographic slope.

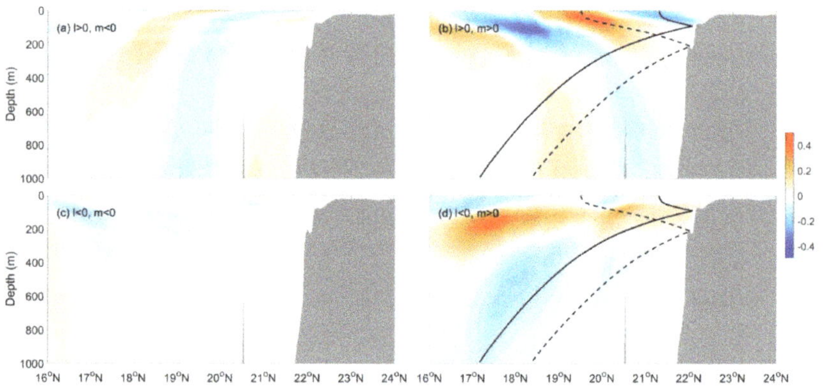

Figure 8. NIWs propagating in different directions along 118° E at 00:00 on 24 October. Note that l and m represent the horizontal and vertical wavenumbers, respectively; $l > 0$ ($l < 0$) corresponds to NIWs propagating northward (southward) and $m < 0$ ($m > 0$) corresponds to NIWs propagating upward (downward). The gray shading in each subfigure indicates the topography. In (**b**,**d**), the black solid and dashed curves represent the ray tracing results.

To study the vertical distribution of Megi-induced NIKE, Figure 9 displays the NIKE at several depths at 00:00 on 24 October as an example. Similar results can be found on the adjacent several days, which are not shown. From Figure 9, we can find that strong NIKE mainly appeared in the upper ocean, especially near the surface. Megi-induced NIKE in the upper 200 m (the upper panels of Figure 9) was at least one order of magnitude stronger than that below 500 m depth (the lower panels of Figure 9). Moreover, there was a slight elevation of NIKE at 1000 m depth, whereas almost no enhancement was found at lower depths. This result suggests that Megi-induced NIWs could only reach an approximate depth of 1000 m in the SCS Basin. In addition, because NIWs are a kind of internal wave which can propagate in both horizontal and vertical directions, the patterns of Megi-induced NIKE at different depths were different.

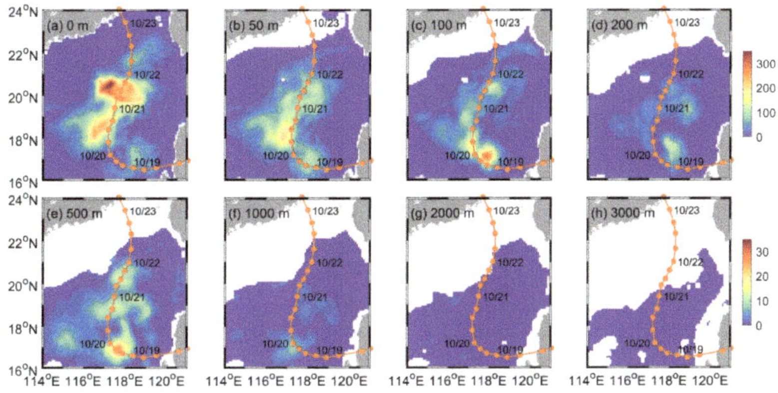

Figure 9. NIKE (shading, unit: J/m^3) at (**a**) 0, (**b**) 50, (**c**) 100, (**d**) 200, (**e**) 500, (**f**) 1000, (**g**) 2000 and (**h**) 3000 m depth at 00:00 on 24 October 2010. Note that the range of colorbar for the upper and lower panels is different. Orange curves denote the track of typhoon Megi.

4.2. Damping

The damping of NIKE is an important characteristic of typhoon-induced NIWs [31,32,34,37]. To investigate the damping of Megi-induced NIWs, Figure 10a illustrates the e-folding time of depth-integrated NIKE. It is interesting to find that the e-folding time of Megi-induced NIWs was site-dependent and varied from several days to approximate a month. Along Megi's track, Megi-induced NIWs damped quickly with the e-folding time generally smaller than one week; whereas away from Megi's track, the e-folding time was longer. It is clearly shown that to the west of Luzon Island and the Luzon Strait, two remarkable zones exist where the e-folding time was longer than 20 days.

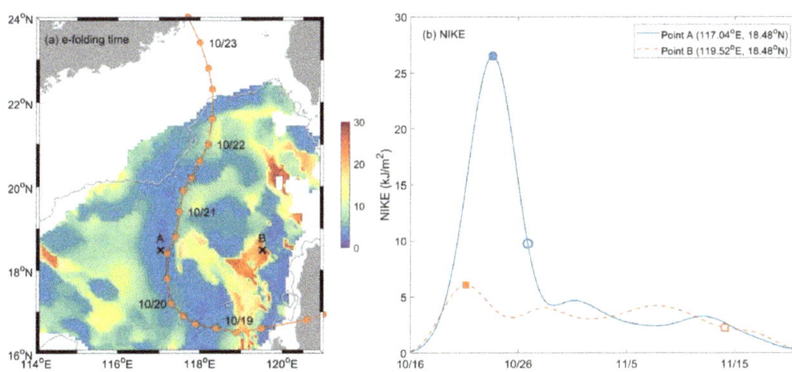

Figure 10. (**a**) The e-folding time (shading, unit: days) of depth-integrated NIKE. The orange curve denotes the track of typhoon Megi. (**b**) Lowpass filtered depth-integrated NIKE at points A (blue solid; 117.04° E, 18.48° N) and B (orange dashed; 119.52° E, 18.48° N) which are marked by black crosses in (**a**). The filled and open circles (squares) locate the maximum and e-folding values of depth-integrated NIKE at point A and B, respectively.

To study the difference of e-folding time along and far away from Megi's track, Figure 10b displays the time series of lowpass filtered depth-integrated NIKE at points A (117.04° E, 18.48° N) and B (119.52° E, 18.48° N) as two examples. At point A which is nearly on Megi's track, the NIKE was quickly strengthened as the response to typhoon Megi, reached the peak value (26.5 kJ/m^2) on 23 October, and quickly damped thereafter.

On 26 October, the NIKE reached the e-folding of its peak value. Therefore, the e-folding time here was 3.1 days. Moreover, the strengthening and damping processes of Megi-induced NIKE at point A were nearly symmetric. However, the situation at point B was much different. First, because point B is far away from Megi's track, the peak value of NIKE was only 6.0 kJ/m^2, which is much smaller than that at point A. Second, the pattern of NIKE evolution here is different from that at point A: Accompanied with the passage of typhoon Megi, the NIKE increased and reached the peak value on 21 October; then the NIKE damped; however, before the NIKE damped to the e-folding of its peak value, it was strengthened again and then exhibited some fluctuations until 7 November. On 14 November, the NIKE at point B reached the e-folding of its peak value. In this case, the e-folding time at point B was 24.0 days.

To explore the possible cause of different damping features of NIKE at points A and B, Figure 11 illustrates the depth-integrated NIKE along the 18.48° N section. As point A is nearly on Megi's track, the NIKE near point A was the strongest along this section. With time going on, the NIKE at point A gradually propagated westward, which is consistent with the result shown in Figure 6. The NIKE at point A damped quickly with significant NIKE existing from 20 to 26 October. As for point B, because it is far away from Megi's track, Megi-induced NIKE here was not significant and only lasted from 19 to 24 October. Thereafter, the NIWs initially generated at 118.8° E propagated eastward to point B and lasted to 1 November, which caused the fluctuations of NIKE at point B (Figure 9c). Similar results can be found at other points with long e-folding time. Based on these results, it can be concluded that the local long damping time of NIKE is related to the NIWs propagating from other sites.

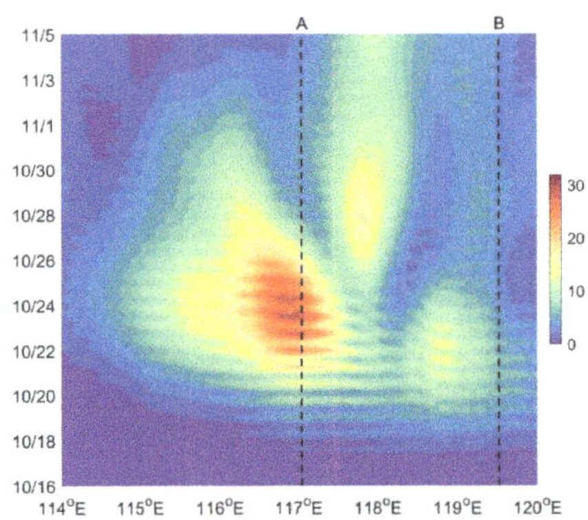

Figure 11. Depth-integrated NIKE (shading, unit: kJ/m^2) as a function of time along 18.48° N. Black dashed lines indicate positions of points A and B.

5. Modal Content

Finally, attention is paid to the modal content of Megi-induced NIWs. We also choose the results at points A and B as examples, for which the modal NIKE is shown in Figure 12. Note that in the modal decomposition in this study, a total of 11 (one barotropic and the first ten baroclinic) modes were taken into consideration.

It is clearly shown that Megi-induced NIWs at point A were dominated by mode-2 which accounted for 45% of the total NIKE. Following mode-2 were mode-3 and mode-1, which occupied 21% and 10% of the total NIKE, respectively. The sum of the first three baroclinic modes accounted for 76% of the total NIKE. The proportions of the other modes in the total NIKE were smaller than 10% and generally exhibited a decreasing trend with the increase in mode number. Moreover, the NIKE of the dominant modes, especially mode-2 and mode-3, synchronously varied with the total NIKE.

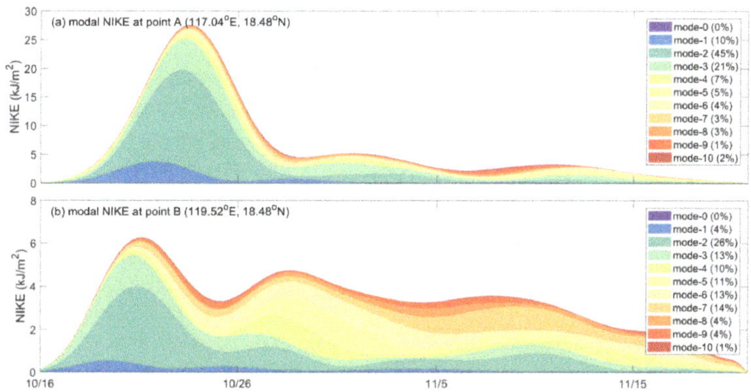

Figure 12. Modal NIKE at points (**a**) A (117.04° E, 18.48° N) and (**b**) B (119.52° E, 18.48° N). The values in the brackets indicate the proportions of modal NIKE in the total NIKE. Note that the modal NIKE has been lowpass filtered to remove oscillations at its own frequency.

The modal content at point B exhibited a different pattern from that at point A. During 16 to 24 October, the locally generated NIWs were dominated by mode-2, mode-3 and mode-1, which was consistent with that at point A; the evolution of mode-2 and mode-3 generally agreed with that of the total NIKE. However, after 24 October, the NIKE of mode-4, mode-5 and mode-6 started to enhance and gradually became dominant. During 5 to 15 November, mode-7 became the most significant mode. On average, mode-2 was the strongest and accounted for 26% of the total NIKE at point B; the contributions of mode-3 to mode-7 were comparable, for which the proportions of modal NIKE in the total NIKE were all greater than 10%.

6. Discussion

The above analysis shows that Megi-induced NIKE was mainly concentrated in the deep SCS basin, which was caused by the reflection of NIWs at the supercritical continental slope around the SCS Basin. A similar phenomenon has been reported for the diurnal internal tides in the SCS [56]. According to the equation of topographic criticality:

$$\gamma = \frac{s_{topo}}{s_{wave}} = \frac{s_{topo}}{\sqrt{(\omega^2 - f^2)/(N^2 - \omega^2)}} \tag{6}$$

where s_{topo} and s_{wave} are the topographic slope and internal wave slope, because the frequency of Megi-induced NIWs (Figure 2) is smaller than that of diurnal internal tides, the s_{wave} for Megi-induced NIWs is smaller than that for diurnal internal tides. In other words, Megi-induced NIWs are more susceptible to reflection on the continental slope in the northern SCS than the diurnal internal tides.

The damping feature of Megi-induced NIWs was site-dependent: In the region near Megi's track, the e-folding time of NIWs was generally less than one week; whereas in two zones to the west of Luzon Island and the Luzon Strait, which are far away from Megi's track, the e-folding time could be longer than 20 days. This result emphasizes the correlation between the distance away from typhoon's track and the decay of NIWs and can partly answer the following questions: Why NIWs generated by different typhoons have different e-folding times at the same position [31] and why NIWs generated by the same typhoon have different e-folding times at different positions [37,57].

The modal decomposition results indicate that Megi-induced NIWs were dominated by the first three baroclinic modes. Along Megi's wake, Megi-induced NIWs quickly dampened after the passage of Megi. However, at point B which is far away from Megi's wake, higher modes (mode-4 to mode-7) appeared and gradually became dominant after

24 October. To explore the possible cause of these higher modes (mode-4 to mode-7), Figure 13 illustrates the zonal currents of NIWs at four points along 18.48° N. By simply counting the times of sign changing of zonal currents in the vertical direction, we can find that these higher modes (mode-4 to mode-7) mainly appeared at point B and 119.04° E, 18.48° N, whereas at point A and 118° E, 18.48° N, these higher modes are nearly invisible. In other words, the appearance of these higher modes (mode-4 to mode-7) was limited to a small region near point B. According to [58], higher modes can be generated when low-mode internal waves interact with mesoscale eddies. Figure 14 illustrates the HYCOM surface elevations from 18 October to 1 November with an interval of 2 days. From Figure 14, we can find that an anticyclonic eddy formed around 24 October and influenced the region around point B. As mentioned above, the NIWs initially generated at 118.8° E propagated eastward to point B after 24 October (Figure 11). This case is similar to that reported by [58]. Therefore, we speculated that the higher modes (mode-4 to mode-7) at point B after 24 October were likely caused by the interaction between NIWs and a mesoscale eddy. However, point A was also under the influence of a cyclonic eddy after 22 October. Why higher modes (mode-4 to mode-7) did not become significant at point A remains unclear. Therefore, the difference of modal content at points A and B still needs further exploration.

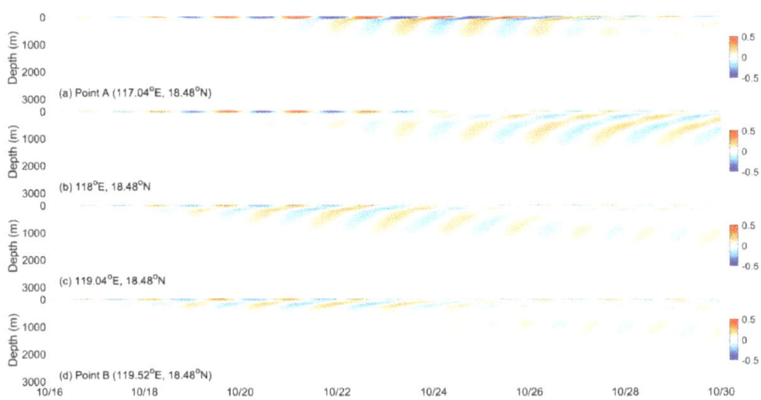

Figure 13. Zonal currents of NIWs (shading, unit: m/s) at (**a**) point A (117.04° E, 18.48° N), (**b**) 118° E, 18.48° N, (**c**) 119.04° E, 18.48° N and (**d**) point B (119.52° E, 18.48° N) from 16 to 30 October 2010.

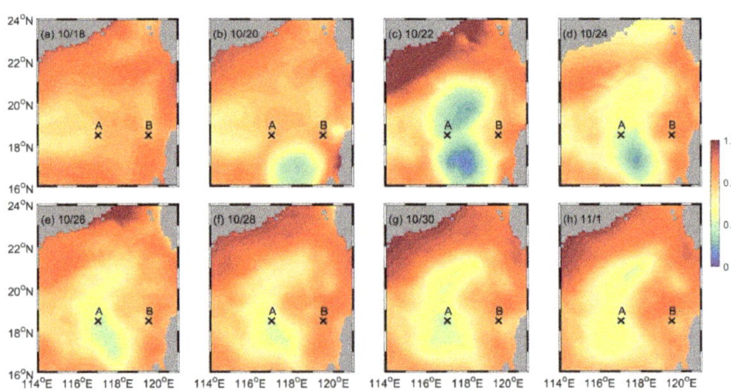

Figure 14. HYCOM surface elevations (shading, unit: m) at 00:00 on (**a**) 18, (**b**) 20, (**c**) 22, (**d**) 24, (**e**) 26, (**f**) 28 and (**g**) 30 October and (**h**) 1 November 2010.

7. Conclusions

Based on the HYCOM reanalysis results, the NIWs generated by typhoon Megi in 2010 in the SCS were investigated in this study. Through a comparison with in situ observations at mooring UIB6, we first showed that the HYCOM reanalysis results can reasonably reproduce typhoon-induced NIWs, which can act as a supplement to in situ observations and provide us with an opportunity to better understand NIWs.

The results indicate that Megi-induced NIWs showed temporal and spatial variations in the SCS. The NIKE in the SCS was rapidly enhancedin response to typhoon Megi. However, the strongest NIKE appeared several days after the passage of Megi, rather than under its influence. Moreover, it is interesting to note that Megi-induced NIKE was mainly concentrated in the deep SCS basin where the water depth is greater than 1000 m, although typhoon Megi passed over both the deep SCS basin and shallow continental shelf and slope in the northern SCS. Through analysis, it was found that the continental slope in the northern SCS is supercritical to Megi-induced NIWs. In other words, when Megi-induced NIWs impinged on the continental slope, they were reflected and trapped in the deep SCS basin. This conclusion was validated by comparing the northward–downward and southward–downward components of Megi-induced NIWs. Moreover, it is found that Megi-induced NIWs could reach 1000 m depth in the vertical direction.

In situ observations have shown that Megi-induced NIWs dampened quickly at mooring UIB6 [32,37]. However, the HYCOM reanalysis results indicate that only in the region near Megi's track did the NIWs exhibit a similar feature, with the e-folding time generally being smaller than one week; whereas in two zones to the west of Luzon Island and the Luzon Strait, which are far away from Megi's track, the e-folding time of Megi-induced NIWs could be longer than 20 days. It was found that the NIWs generated at other sites could propagate to the two zones and influence the local NIKE, which accounted for the long e-folding time.

The modal content of Megi-induced NIWs was also explored in this study. Near Megi's track, Megi-induced NIWs were dominated by mode-2, which was followed by mode-3 and mode-1. The three modes accounted for 76% of the total NIKE. However, at the region far away from Megi's track, the NIKE of higher modes (mode-4 to mode-7) was enhanced several days after the passage of typhoon Megi. The cause of these higher modes still needs further exploration.

Author Contributions: Conceptualization, A.C.; methodology, A.C. and Z.G.; formal analysis, A.C., Z.G. and Y.P.; writing—original draft preparation, A.C.; writing—review and editing, A.C., Z.G., J.S., H.H. and P.L.; supervision, A.C. and J.S.; funding acquisition, A.C. and H.H. All authors have read and agreed to the published version of the manuscript.

Funding: This research was funded by the Natural Science Foundation of Zhejiang Province through grant LY21D060005, the Joint Project of Zhoushan Municipality and Zhejiang University through grant 2019C81060, and the National Natural Science Foundation of China through grant 41621064. This research was also supported by the Laboratory for Regional Oceanography and Numerical Modeling, Qingdao National Laboratory for Marine Science and Technology through grant 2019A01.

Conflicts of Interest: The authors declare no conflict of interest.

References

1. Alford, M.H.; MacKinnon, J.A.; Simmons, H.L.; Nash, J.D. Near-Inertial Internal Gravity Waves in the Ocean. *Annu. Rev. Mar. Sci.* **2016**, *8*, 95–123. [CrossRef]
2. Garrett, C. Mixing with latitude. *Nat. Cell Biol.* **2003**, *422*, 477. [CrossRef] [PubMed]
3. Carter, G.; Fringer, O.; Zaron, E. Regional Models of Internal Tides. *Oceanography* **2012**, *25*, 56–65. [CrossRef]
4. Zhang, Z.; Qiu, B.; Tian, J.; Zhao, W.; Huang, X. Latitude-dependent finescale turbulent shear generations in the Pacific trop-ical-extratropical upper ocean. *Nat. Commun.* **2018**, *9*, 4086. [CrossRef] [PubMed]
5. Cao, A.; Guo, Z.; Wang, S.; Chen, X.; Lv, X.; Song, J. Upper ocean shear in the northern South China Sea. *J. Oceanogr.* **2019**, *75*, 525–539. [CrossRef]
6. Jing, Z.; Wu, L. Low-Frequency Modulation of Turbulent Diapycnal Mixing by Anticyclonic Eddies Inferred from the HOT Time Series. *J. Phys. Oceanogr.* **2013**, *43*, 824–835. [CrossRef]

7. Whalen, C.B.; MacKinnon, J.A.; Talley, L.D. Large-scale impacts of the mesoscale environment on mixing from wind-driven internal waves. *Nat. Geosci.* **2018**, *11*, 842–847. [CrossRef]
8. Alford, M.H. Internal Swell Generation: The Spatial Distribution of Energy Flux from the Wind to Mixed Layer Near-Inertial Motions. *J. Phys. Oceanogr.* **2001**, *31*, 2359–2368. [CrossRef]
9. Alford, M.H. Improved global maps and 54-year history of wind-work on ocean inertial motions. *Geophys. Res. Lett.* **2003**, *30*, 1424. [CrossRef]
10. Watanabe, M.; Hibiya, T. Global estimates of the wind-induced energy flux to inertial motions in the surface mixed layer. *Geophys. Res. Lett.* **2002**, *29*, 64-1–64-3. [CrossRef]
11. Furuichi, N.; Hibiya, T.; Niwa, Y. Model-predicted distribution of wind-induced internal wave energy in the world's oceans. *J. Geophys. Res. Oceans* **2008**, *113*, C09034. [CrossRef]
12. Jiang, J.; Lu, Y.; Perrie, W. Estimating the energy flux from the wind to ocean inertial motions: The sensitivity to surface wind fields. *Geophys. Res. Lett.* **2005**, *32*, 291–310. [CrossRef]
13. Rimac, A.; Storch, J.; Eden, C.; Haak, H. The influence of high-resolution wind stress field on the power input to near-inertial motions in the ocean. *Geophys. Res. Lett.* **2013**, *40*, 4882–4886. [CrossRef]
14. Simmons, H.; Alford, M. Simulating the Long-Range Swell of Internal Waves Generated by Ocean Storms. *Oceanography* **2012**, *25*, 30–41. [CrossRef]
15. Munk, W.; Wunsch, C. Abyssal recipes II: Energetics of tidal and wind mixing. *Deep. Sea Res. Part I Oceanogr. Res. Pap.* **1998**, *45*, 1977–2010. [CrossRef]
16. Egbert, G.D.; Ray, R.D. Significant dissipation of tidal energy in the deep ocean inferred from satellite altimeter data. *Nat. Cell Biol.* **2000**, *405*, 775–778. [CrossRef]
17. Alford, M.H.; MacKinnon, J.A.; Zhao, Z.; Pinkel, R.; Klymak, J.; Peacock, T. Internal waves across the Pacific. *Geophys. Res. Lett.* **2007**, *34*, 24601. [CrossRef]
18. Xie, X.-H.; Shang, X.-D.; Chen, G.-Y.; Sun, L. Variations of diurnal and inertial spectral peaks near the bi-diurnal critical latitude. *Geophys. Res. Lett.* **2009**, *36*, 02606. [CrossRef]
19. MacKinnon, J.A.; Alford, M.H.; Sun, O.; Pinkel, R.; Zhao, Z.; Klymak, J. Parametric Subharmonic Instability of the Internal Tide at 29° N. *J. Phys. Oceanogr.* **2013**, *43*, 17–28. [CrossRef]
20. Nikurashin, M.; Legg, S. A Mechanism for Local Dissipation of Internal Tides Generated at Rough Topography. *J. Phys. Oceanogr.* **2011**, *41*, 378–395. [CrossRef]
21. Liang, X.; Wunsch, C. Note on the redistribution and dissipation of tidal energy over mid-ocean ridges. *Tellus A Dyn. Meteorol. Oceanogr.* **2015**, *67*, 27385. [CrossRef]
22. Wang, S.; Cao, A.; Chen, X.; Li, Q.; Song, J. On the resonant triad interaction over mid-ocean ridges. *Ocean. Model.* **2021**, *158*, 101734. [CrossRef]
23. Nikurashin, M.; Ferrari, R. Radiation and dissipation of internal waves generated by geostrophic gotions impinging on small-scale topography: Theory. *J. PhysOceanogr.* **2010**, *40*, 2025–2042.
24. Alford, M.H.; Shcherbina, A.Y.; Gregg, M.C. Observations of Near-Inertial Internal Gravity Waves Radiating from a Frontal Jet. *J. Phys. Oceanogr.* **2013**, *43*, 1225–1239. [CrossRef]
25. Liang, X.; Thurnherr, A.M. Eddy-Modulated Internal Waves and Mixing on a Midocean Ridge. *J. Phys. Oceanogr.* **2012**, *42*, 1242–1248. [CrossRef]
26. D'Asaro, E.; Black, P.; Centurioni, L.; Harr, P.; Jayne, S.; Lin, I.; Lee, C.; Morzel, J.; Mrvaljevic, R.; Niiler, P.; et al. Typhoon-Ocean Interaction in the Western North Pacific: Part 1. *Oceanography* **2011**, *24*, 24–31. [CrossRef]
27. Nguyen, C.H.; Owen, J.S.; Franke, J.; Neves, L.C.; Hargreaves, D.M. Typhoon track simulations in the North West Pacific: Informing a new wind map for Vietnam. *J. Wind. Eng. Ind. Aerodyn.* **2021**, *208*, 104441. [CrossRef]
28. Wang, L.; Li, J.; Lizhen, W.; Jiachun, L. Estimation of extreme wind speed in SCS and NWP by a non-stationary model. *Theor. Appl. Mech. Lett.* **2016**, *6*, 131–138. [CrossRef]
29. Li, J.; Xu, J.; Liu, J.; He, Y.; Chen, Z.; Cai, S. Correlation of Near-Inertial Wind Stress in Typhoon and Typhoon-Induced Oceanic Near-Inertial Kinetic Energy in the Upper South China Sea. *Atmosphere* **2019**, *10*, 388. [CrossRef]
30. Sun, Z.; Hu, J.; Zheng, Q.; Li, C. Strong near-inertial oscillations in geostrophic shear in the northern South China Sea. *J. Oceanogr.* **2011**, *67*, 377–384. [CrossRef]
31. Chen, G.; Xue, H.; Wang, D.; Xie, Q. Observed near-inertial kinetic energy in the northwestern South China Sea. *J. Geophys. Res. Oceans* **2013**, *118*, 4965–4977. [CrossRef]
32. Guan, S.; Zhao, W.; Huthnance, J.; Tian, J.; Wang, J. Observed upper ocean response to typhoon Megi (2010) in the Northern South China Sea. *J. Geophys. Res. Oceans* **2014**, *119*, 3134–3157. [CrossRef]
33. Yang, B.; Hou, Y. Near-inertial waves in the wake of 2011 Typhoon Nesat in the northern South China Sea. *Acta Oceanol. Sin.* **2014**, *33*, 102–111. [CrossRef]
34. Yang, B.; Hou, Y.; Hu, P.; Liu, Z.; Liu, Y. Shallow ocean response to tropical cyclones observed on the continental shelf of the northwestern South China Sea. *J. Geophys. Res. Oceans* **2015**, *120*, 3817–3836. [CrossRef]
35. Zhang, H.; Chen, D.; Zhou, L.; Liu, X.; Ding, T.; Zhou, B. Upper ocean response to typhoon Kalmaegi (2014). *J. Geophys. Res. Oceans* **2016**, *121*, 6520–6535. [CrossRef]

36. Zhang, H.; Wu, R.; Chen, D.; Liu, X.; He, H.; Tang, Y.; Ke, D.; Shen, Z.; Li, J.; Xie, J.; et al. Net Modulation of Upper Ocean Thermal Structure by Typhoon Kalmaegi (2014). *J. Geophys. Res. Oceans* **2018**, *123*, 7154–7171. [CrossRef]
37. Cao, A.; Guo, Z.; Song, J.; Lv, X.; He, H.; Fan, W. Near-Inertial Waves and Their Underlying Mechanisms Based on the South China Sea Internal Wave Experiment (2010–2011). *J. Geophys. Res. Oceans* **2018**, *123*, 5026–5040. [CrossRef]
38. Ding, W.; Liang, C.; Liao, G.; Li, J.; Lin, F.; Jin, W.; Zhu, L. Propagation characteristics of near-inertial waves along the conti-nental shelf in the wake of the 2008 Typhoon Hagupit in the northern South China Sea. *B. Mar. Sci.* **2018**, *94*, 1293–1311. [CrossRef]
39. Ma, Y.; Zhang, S.; Qi, Y.; Jing, Z. Upper ocean near-inertial response to the passage of two sequential typhoons in the north-western South China Sea. *Sci. China Earth Sci.* **2019**, *62*, 863–871. [CrossRef]
40. Xu, J.; Huang, Y.; Chen, Z.; Liu, J.; Liu, T.; Li, J.; Cai, S.; Ning, D. Horizontal variations of typhoon-forced near-inertial oscillations in the south China sea simulated by a numerical model. *Cont. Shelf Res.* **2019**, *180*, 24–34. [CrossRef]
41. Ying, M.; Zhang, W.; Yu, H.; Lu, X.; Feng, J.; Fan, Y.; Zhu, Y.; Chen, D. An Overview of the China Meteorological Administration Tropical Cyclone Database. *J. Atmos. Ocean. Technol.* **2014**, *31*, 287–301. [CrossRef]
42. Mercier, M.J.; Garnier, N.B.; Dauxois, T. Reflexion and Diffraction of Internal Waves analyzed with the Hilbert Transform. *Phys. Fluids* **2008**, *20*, 086601. [CrossRef]
43. Wang, S.; Cao, A.; Chen, X.; Li, Q.; Song, J.; Meng, J. Estimation of the Reflection of Internal Tides on a Slope. *J. Ocean. Univ. China* **2020**, *19*, 489–496. [CrossRef]
44. Nash, J.D.; Alford, M.H.; Kunze, E. Estimating Internal Wave Energy Fluxes in the Ocean. *J. Atmos. Ocean. Technol.* **2005**, *22*, 1551–1570. [CrossRef]
45. Zhao, Z.; Alford, M.H.; Mackinnon, J.A.; Pinkel, R. Long-Range Propagation of the Semidiurnal Internal Tide from the Ha-waiian Ridge. *J. Phys. Oceanogr.* **2010**, *40*, 713–736. [CrossRef]
46. Cao, A.-Z.; Li, B.-T.; Lv, X.-Q. Extraction of Internal Tidal Currents and Reconstruction of Full-Depth Tidal Currents from Mooring Observations. *J. Atmos. Ocean. Technol.* **2015**, *32*, 1414–1424. [CrossRef]
47. Shang, X.; Liu, Q.; Xie, X.; Chen, G.; Chen, R. Characteristics and seasonal variability of internal tides in the southern South China Sea. *Deep. Sea Res. Part I Oceanogr. Res. Pap.* **2015**, *98*, 43–52. [CrossRef]
48. Zhao, Z.; Alford, M.H.; Lien, R.-C.; Gregg, M.C.; Carter, G.S. Internal Tides and Mixing in a Submarine Canyon with Time-Varying Stratification. *J. Phys. Oceanogr.* **2012**, *42*, 2121–2142. [CrossRef]
49. D'Asaro, E.A.; Eriksen, C.C.; Levine, M.D.; Niiler, P.; Paulson, C.A.; Meurs, P.V. Upper-ocean inertial currents forced by a strong storm. Part I: Data and comparisons with linear theory. *J. Phys. Oceanogr.* **1995**, *25*, 2909–2936. [CrossRef]
50. Gill, A.E. On the Behavior of Internal Waves in the Wakes of Storms. *J. Phys. Oceanogr.* **1984**, *14*, 1129–1151. [CrossRef]
51. Sun, O.M.; Pinkel, R. Energy Transfer from High-Shear, Low-Frequency Internal Waves to High-Frequency Waves near Kaena Ridge, Hawaii. *J. Phys. Oceanogr.* **2012**, *42*, 1524–1547. [CrossRef]
52. Guo, Z.; Cao, A.; Lv, X.; Song, J. Impact of multiple tidal forcing on the simulation of the M2 internal tides in the northern South China Sea. *Ocean.Dyn.* **2019**, *70*, 187–198. [CrossRef]
53. Price, J.F.; Sanford, T.B.; Forristall, G.Z. Forced Stage Response to a Moving Hurricane. *J. Phys. Oceanogr.* **1994**, *24*, 233–260. [CrossRef]
54. Cacchione, D.A.; Pratson, L.F.; Ogston, A.S. The shaping of continental slopes by internal tides. *Science* **2020**, *296*, 724–727. [CrossRef] [PubMed]
55. Guo, Z.; Cao, A.; Lv, X.; Song, J. Impacts of Stratification Variation on the M2 Internal Tide Generation in Luzon Strait. *Atmos. Ocean.* **2020**, *58*, 206–218. [CrossRef]
56. Wu, L.; Miao, C.; Zhao, W. Patterns of K1 and M2 internal tides and their seasonal variations in the northern South China Sea. *J. Oceanogr.* **2013**, *69*, 481–494. [CrossRef]
57. Hu, S.; Liu, L.; Guan, C.; Zhang, L.; Hu, D. Dynamic features of near-inertial oscillations in the Northwest Pacific derived from mooring observations from 2015 to 2018. *J. Oceanol. Limn.* **2020**, *38*, 1092–1107. [CrossRef]
58. Dunphy, M.; Lamb, K.G. Focusing and vertical mode scattering of the first mode internal tide by mesoscale eddy interaction. *J. Geophys. Res. Oceans* **2014**, *119*, 523–536. [CrossRef]

Journal of *Marine Science and Engineering*

Article

Distribution and Source Sites of Nonlinear Internal Waves Northeast of Hainan Island

Jianjun Liang [1,2,*], Xiao-Ming Li [1,2] and Kaiguo Fan [3]

1 Key Laboratory of Earth Observation of Hainan Province, Hainan Research Institute, Aerospace Information Research Institute, Chinese Academy of Sciences, Sanya 572000, China; lixm@radi.ac.cn
2 Key Laboratory of Digital Earth Science, Aerospace Information Research Institute, Chinese Academy of Sciences, Beijing 100094, China
3 School of Meteorology and Oceanography, National University of Defense Technology, Changsha 410015, China; van.fkg@tom.com
* Correspondence: liangjj@radi.ac.cn

Abstract: The distribution and source sites of nonlinear internal waves (NLIWs) northeast of Hainan Island were investigated using satellite observations and a wavefront propagation model. Satellite observations show two types of NLIWs (here referred to as type-S and type-D waves). The type-S waves are spaced at a semidiurnal tidal period and the type-D waves are spaced at a diurnal tidal period. The spatial distribution of the two types of NLIWs displays a sandwich structure in which the middle region is influenced by both types of NLIWs, and the northern and southern regions are governed by the type-S and type-D waves, respectively. Solving the wavefront model yields good agreement between simulated and observed wavefronts from the Luzon Strait to Hainan Island. We conclude that the NLIWs originate from the Luzon Strait.

Keywords: internal waves; South China Sea; Hainan Island

Citation: Liang, J.; Li, X.-M.; Fan, K. Distribution and Source Sites of Nonlinear Internal Waves Northeast of Hainan Island. *J. Mar. Sci. Eng.* **2022**, *10*, 55. https://doi.org/10.3390/jmse10010055

Academic Editors: SungHyun Nam and Xueen Chen

Received: 22 November 2021
Accepted: 30 December 2021
Published: 4 January 2022

1. Introduction

Nonlinear internal waves (NLIWs) are often transformed from the propagation of nonlinear internal tides. They frequently occur in coastal oceans and marginal seas, and have strong effects on driving vertical mixing, scattering and ducting acoustic modes, and endangering underwater equipment and moving objects. Therefore, it is very important to know their source sites and to evaluate these effects.

The NLIWs in the northern South China Sea (SCS) have received a great deal of attention, particularly those from the Luzon Strait to Dongsha Atoll [1]. The Luzon Strait radiates strong semidiurnal and diurnal internal tides westward into the SCS due to the interactions of strong barotropic tidal currents [2] with the two shallow ridges: the Heng-Chun ridge and the Lan-Yu ridge [3]. In the deep basin, the semidiurnal internal tides can steepen to form NLIWs under the influence of nonhydrostatic and rotational dispersion; by contrast, the rotation significantly inhibits the steepening of diurnal internal tides [4]. Then, the NLIWs diffract and refract around Dongsha Atoll [5], dissipating most of their energy [6,7]. Continuing with northwestward propagation, the NLIWs may experience a polarity conversion under the condition that the pycnocline is below the mid-depth [8].

Satellite observations show that NLIWs also occur frequently on the continental slope-shelf northeast of Hainan Island [9,10], as shown by the black rectangle in Figure 1. Moreover, field measurements show that the largest NLIWs in the region can reach an amplitude of 45 m at a water depth of 117 m [11], making them among the strongest waves observed on global continental shelves [12].

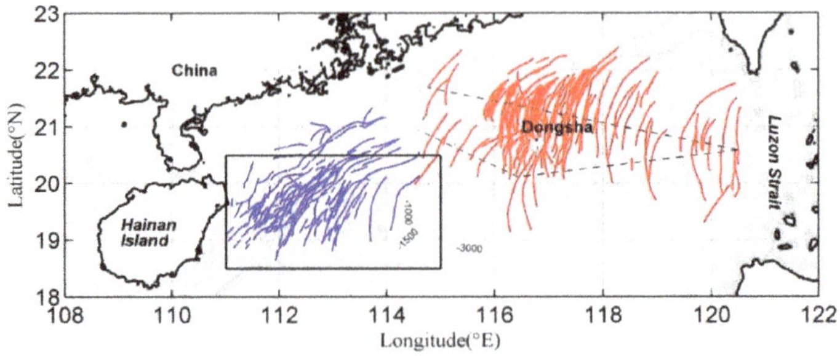

Figure 1. Satellite observations of NLIWs. The black rectangle shows the study region. The red and black lines represent the satellite-observed wavefronts of NLIWs. The red lines are those provided by Zhao et al. [13], and the blue lines are extracted from the collected satellite images in this study. The black dashed lines show the commonly recognized propagation path of NLIWs originating from the Luzon Strait.

By calculating barotropic tidal forcing and analyzing three Envisat Advanced Synthetic Aperture Radar (ASAR) images, Li et al. [14] suggests that the SAR-observed NLIWs northeast of Hainan Island originate from the Luzon Strait. A common view of the NLIWs originating from the Luzon Strait is that they can transit through the deep basin and diffract around Dongsha Atoll, as shown by the black dashed lines in Figure 1 [5,15]. However, whether these NLIWs can directly arrive at the area northeast of Hainan Island remains unclear. In contrast, Xu et al. [16] found that the field-observed NLIWs are generated by nonlinear transformation of diurnal internal tides and they suggest that the diurnal internal tides arise from tide-topography interactions at the local continental shelf-break, distinct from the result of Li et al. [14]. The two above-mentioned opposing opinions on the NLIWs northeast of Hainan Island warrant further investigations into their source sites.

Moreover, apart from the NLIWs separated by a diurnal tidal period as observed by Xu et al. [16], we also found NLIWs separated by a semidiurnal tidal period by analyzing the satellite data. Hence, there are two types of NLIWs northeast of Hainan Island, referred to as type-S (S denotes semidiurnal) and type-D (D denotes diurnal) waves. A meaningful question then arises of whether there is a general distribution law for these two types of NLIWs. Here, we combine satellite observations and a wavefront propagation model to clarify the distribution and source sites of NLIWs northeast of Hainan Island.

2. Materials and Methods

2.1. Satellite Data

We collected 22 Envisat ASAR images in the three years 2005, 2011, and 2012 to show the spatial distribution of NLIWs. All the SAR images are preprocessed by radiometric calibration, Lee filtering, and geolocation. The leading wave in a NLIWs packet is depicted as a blue line in Figure 1. The SAR observations clearly show complete type-S waves and only show some segments of type-D waves. Hence, one Moderate Resolution Imaging Spectrometer (MODIS) image acquired on 10 September 2005 was used to clearly show a pair of complete type-D waves.

2.2. A Wavefront Propagation Model

We used the Eikonal equation [17] to simulate the propagation path of NLIWs because the waves always appear as long stripes (referred to as wavefronts) in satellite images. The equation is:

$$T_x^2 + T_y^2 = 1/c(x,y)^2\,, \tag{1}$$

where T is the time for a wave propagating from an origin to a location (x, y) and $c(x, y)$ is the local wave speed. Here, we used the statistical propagation speed model given by

Jackson [18] to calculate $c(x, y)$. The model relates the wave speed in the northern SCS to local water depth $H(x, y)$ in the following manner:

$$c(x,y) = 2.971 \times \sqrt{\tan\mathrm{h}(0.003 + H(x,y)/1390.758)}. \quad (2)$$

Thus, the propagation path of NLIWs can be determined when the origin is known. According to the simulated and measured baroclinic energy flux in the Luzon Strait, we specified the origin near the Batan Islands [19]. Equation (1) was solved by the fast marching method [20].

3. Results

3.1. Distribution of the Two Types of NLIWs

An Enivsat ASAR image acquired on 8 June 2011 presenting typical type-S waves is shown in Figure 2. There are four NLIWs with long wavefronts labeled S1, S2, S3, and S4. The separation distances between S1, S2, S3, and S4 are 92, 59, and 48 km, respectively. According to Equation (2), the decrease in the separation distance is caused by a decrease in wave speed from 2.34 m/s at 1000 m through 1.75 m/s at 500 m to 0.81 m/s at 100 m. Comparing the four NLIWs with simulated wavefronts, we found the interval between them is approximately a semidiurnal tidal period. Because it takes 82 h for a wave to propagate from the Luzon Strait to the position of S1, we examined the tidal forcing during 3–5 June in the Luzon Strait. The TPXO7.2 tidal model [21] predicts the semidiurnal barotropic tides are near the semidiurnal spring tide, indicating the strong generation of semidiurnal internal tides in the Luzon Strait. Thus, the four NLIWs are probably generated by the nonlinear transformation of four successive semidiurnal internal tides, leading to a semidiurnal tidal period interval. In addition, there is a NLIWs with a shorter wavefront, labeled SA1, and the wavefront breaks up into two arms near the Shenhu Ansha Shoal, located at approximately 19.5° N, 112.9° E. The SA1 was generated by the diffraction of S3 at a time earlier than the SAR acquisition time. This diffraction process is the same as the well-known one near Dongsha Atoll [14].

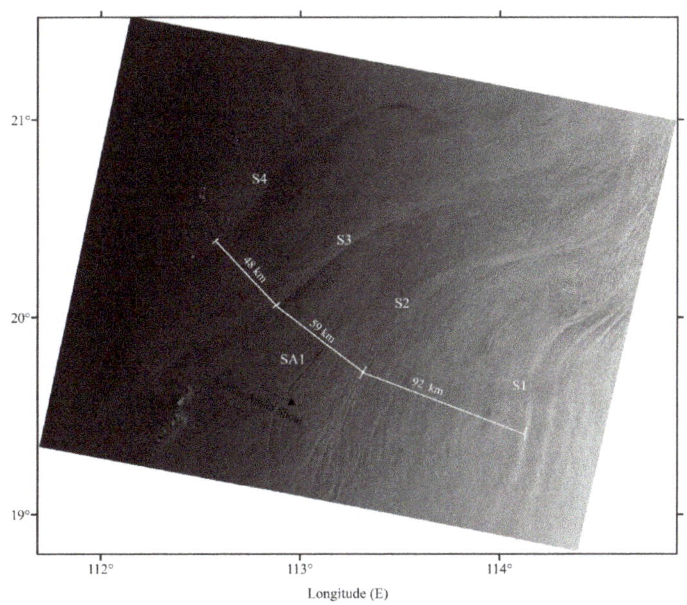

Figure 2. An Envisat ASAR image acquired at 02:30:04 UTC on 8 June 2011.

A representative satellite observation of the type-D waves by MODIS is shown in Figure 3. There are two wave packets labeled D1 and D2. The distance between D1 and D2 is 97 km, and the associated mean wave speed is 1.03 m/s. A comparison between the two wave packets and simulated wavefronts clearly reveals that the interval between them (approximately 26 h) is close to a diurnal tidal period. In addition, the satellite observation agrees with the field measurements, which also show that the NLIWs are separated by a diurnal tidal period [16]. However, the TPXO7.2 tidal model predicts the barotropic diurnal tides during 5–7 September in the Luzon Strait are near diurnal neap tide (close to zero on 7 September), suggesting weak generation of diurnal internal tides. By comparison, the barotropic semidiurnal tides are still near semidiurnal spring tide, resulting in strong generation of semidiurnal internal tides. The weak generation of diurnal internal tides and strong generation of semidiurnal internal tides appear to contradict the fact that D1 and D2 are separated by a diurnal tidal period. A hypothesis is that the parametric subharmonic instability of semidiurnal internal tides may lead to the generation of a near-diurnal internal wave on the continental slope [22] or in the deep basin [23] in the northern SCS, which accounts for the diurnal tidal period interval between D1 and D2.

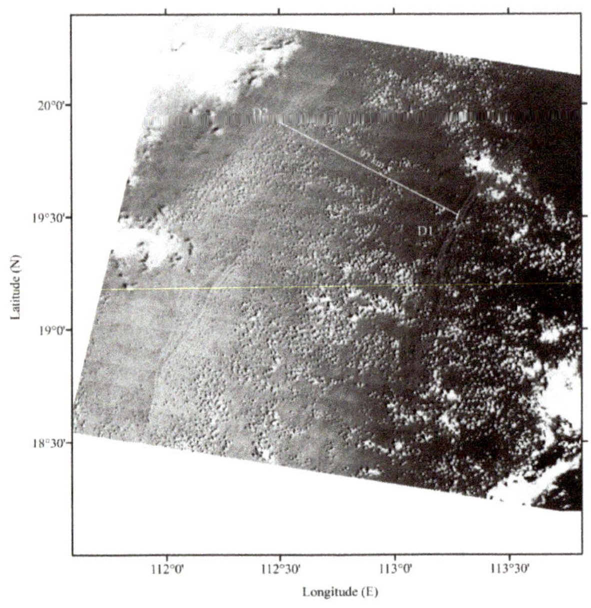

Figure 3. A MODIS image acquired at 03:20:00 UTC on 10 September 2005.

As shown in Figure 1, the wave patterns illustrated by the blue lines are very complicated. Despite the complication, we found a general distribution law for the two types of NLIWs by comparing other satellite observations with the two cases shown in Figures 2 and 3. Basically, the wave processes observed in other satellite images can be explained as variations in the two cases. This argument is supported by the inherent nature of the tidal origin for the NLIWs. Thus, we plotted all the type-S and type-D waves as blue and magenta lines, respectively, in Figure 4. The boundaries separating the two types of NLIWs were drawn by visual inspection of their spatial distribution. As can be seen, the region labeled RS (R denotes region) is dominated by type-S waves, the region labeled RSD is influenced by both types of NLIWs, and the region labeled RD is dominated by type-D waves. Note that the region RD contains the mooring position of Xu et al. [16], which sees remarkable NLIWs separated by a diurnal tidal period.

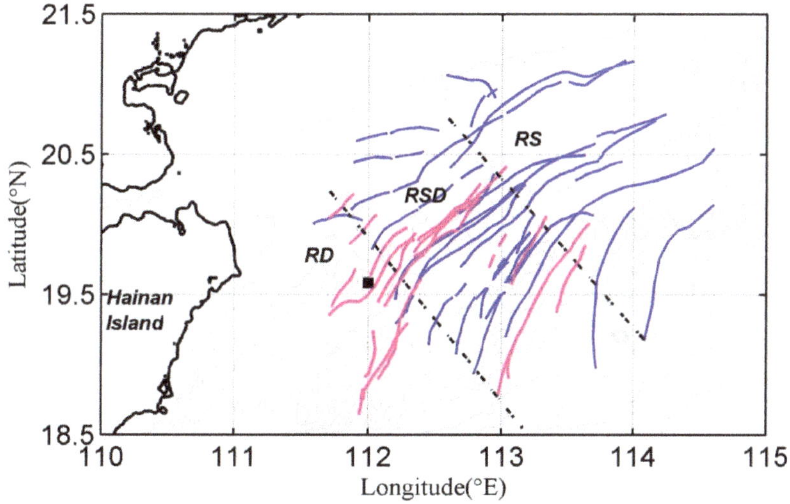

Figure 4. A general distribution pattern of the NLIWs. The blue lines represent type-S waves and the magenta lines represent type-D waves. The two dashed lines depict the boundaries for the distribution of the two types of NLIWs. The filled black square denotes the mooring position of Xu et al. [16].

3.2. Source Sites of the Two Types of NLIWs

Following the distinction of the two types of NLIWs, we next examine their source sites. Solving Equation (1) obtains the simulated wave propagation from the Luzon Strait to Hainan Island (black lines in Figure 5). The simulated wavefronts agree with all the observed wavefronts, including those from the deep basin of the SCS through Dongsha Atoll and the continental slope-shelf to the coastal region of Hainan Island. In particular, the simulated wavefronts agree surprisingly well with the easternmost wavefront and northern portions of other wavefronts in Figure 6a and the two wavefronts in Figure 6b. The agreement shown in Figures 5 and 6 indicates that the source site of NLIWs northeast of Hainan Island is in the Luzon Strait. Further evidence that supports this argument is that both the easternmost wavefronts in Figure 6a,b appear on the continental slope and run across the water depths from 200 to 1000 m, suggesting they have a remote source site.

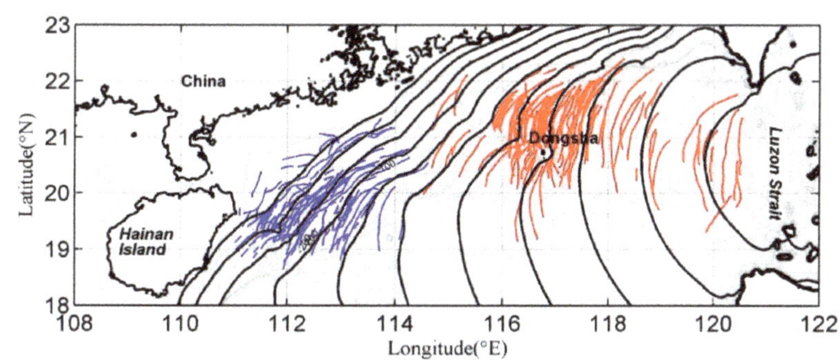

Figure 5. Distribution of simulated wavefronts. The black lines are drawn every 12 h.

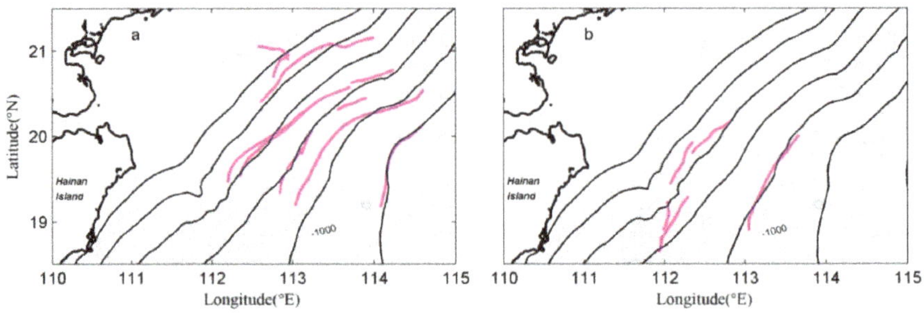

Figure 6. Specific comparisons for simulated and observed wavefronts for the type-S waves (**a**) and type-D waves (**b**). The magenta lines in (**a**,**b**) are extracted from Figures 2 and 3, respectively. The black lines are drawn every 12 h.

4. Discussion and Conclusions

Li et al. [14] found that the NLIWs also originate from the Luzon strait by analyzing three Envisat ASAR images and the phase-locked relation between the NLIWs and the tidal forcing in the Luzon Strait. However, they did not further discriminate the NLIWs separated by a semidiurnal and diurnal tidal period. Li et al. [14] labeled four NLIWs S1, S2, S3, and S4 in the SAR image taken on 6 November 2005 (Figure 7). They noted that S1 and S3 are generated with a 36 h lag in the Luzon Strait, implying that they are generated within two lunar days. However, we found that S1 and S3 have a 12 h lag by comparing the observed and simulated wavefronts. The generation time lag of 36 h may not be reduced to 12 h when they propagate from the Luzon Strait to the SAR observation site. In addition, analysis of the barotropic tidal forcing during 1–3 November in the Luzon Strait reveals that semidiurnal tides are near semidiurnal spring tide, and diurnal tides are in the transition period from diurnal neap to spring tide (Figure 8). The tidal forcing implies the generation of strong semidiurnal internal tides and somewhat weak diurnal internal tides. Thus, a more reasonable explanation is that S1 and S3 are generated by successive semidiurnal internal tides within one lunar day (Figure 8). Given that the crest length of S2 wavefront is significantly shorter than those of S1 and S3, and S2 is more southerly, we infer that S2 is generated by the nonlinear transformation of the diurnal internal tide formed within the same lunar day as the semidiurnal internal tides, according to Figures 4 and 8.

Although the wavefront propagation model developed by Jackson [18] is sufficiently robust to simulate the propagation of wavefronts from the Luzon Strait to continental slopes and Dongsha Atoll [24], it has relatively large uncertainties in predicting the propagation of wavefronts near shelf-break regions, particularly as revealed by the second easternmost wavefront in Figure 6a. This mismatch is caused by the effects of wave amplitudes and mesoscale currents on the propagation speed, which are not well accounted for in the statistical propagation model proposed by Jackson [18]. A more accurate propagation speed model for shelf-break regions needs to be developed.

As revealed by Figures 2, 3 and 7, the easternmost NLIWs will undergo complicated evolution when they propagate onshore, such as wave fission, polarity transformation, refraction, and dissipation, leading to more dense wavefronts on the middle continental shelf. The wave evolution is probably caused by the highly variable continental slope-shelf topography and mesoscale eddies and fronts. How and where these processes primarily affect the vertical mixing, local ecosystem, and sediment transport remains unknown. These questions may need further investigations because the NLIWs are among the largest waves on the global continental shelves.

The confusion regarding the source sites of NLIWs northeast of Hainan Island has persisted for nearly ten years. Here, we combine satellite observations and a wavefront propagation model to clarify the two types of NLIWs, provide a general law for their distribution, and demonstrate that these waves originate from the Luzon Strait.

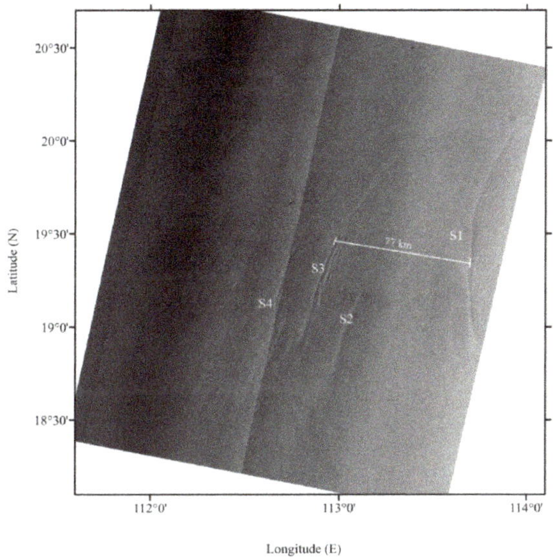

Figure 7. An Envisat ASAR image acquired at 02:28:22 UTC on 6 November 2005.

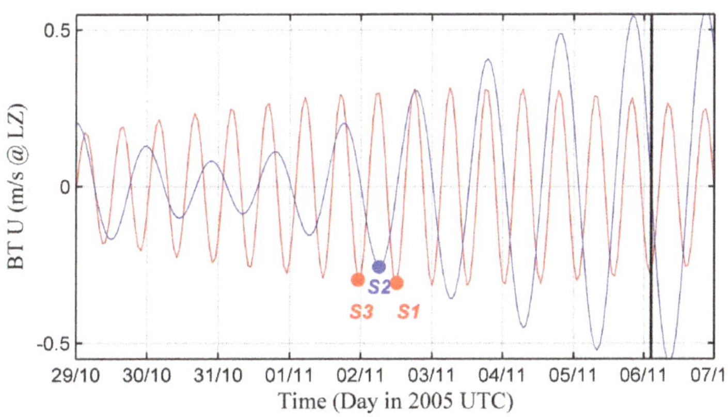

Figure 8. Semidiurnal (red) and diurnal (blue) components of zonal tidal current predictions (TPXO 7.2) in the Luzon Strait. The dots S1, S2, and S3 indicate the westward tidal current peaks corresponding to the S1, S2, and S3 NLIWs in Figure 7, respectively. The thick black line indicates the SAR imaging time in Figure 7.

Author Contributions: Conceptualization, J.L. and X.-M.L.; Data curation, J.L.; Formal analysis, J.L.; Funding acquisition, J.L.; Investigation, J.L.; Methodology, J.L.; Project administration, J.L.; Resources, J.L.; Software, J.L.; Supervision, J.L.; Validation, J.L., X.-M.L. and K.F.; Visualization, J.L. and X.-M.L.; Writing—original draft, J.L.; Writing—review & editing, J.L., X.-M.L. and K.F. All authors have read and agreed to the published version of the manuscript.

Funding: This research was funded by Hainan Provincial Natural Science Foundation of China, grant number 420MS073.

Institutional Review Board Statement: Not applicable.

Informed Consent Statement: Not applicable.

Data Availability Statement: The SAR images were downloaded from http://ids.ceode.ac.cn/Index.aspx, accessed on 21 November 2021, and the MODIS image was downloaded from https://ladsweb.modaps.eosdis.nasa.gov, accessed on 21 November 2021. The red lines in Figure 1 were downloaded from https://figshare.com/authors/Zhongxiang_Zhao/5776886, accessed on 21 November 2021.

Acknowledgments: We thank the Open Spatial Data Sharing Project launched by the Aerospace Information Research Institute for providing the SAR data and NASA's LAADS Web for providing the MODIS image.

Conflicts of Interest: The authors declare no conflict of interest.

References

1. Guo, C.C.; Chen, X.E. A review of internal solitary wave dynamics in the northern South China Sea. *Prog. Oceanogr.* **2014**, *121*, 7–23. [CrossRef]
2. Ye, A.L.; Robinson, I.S. Tidal dynamics in the South China Sea. *Geophys. J. R. Astr. Soc.* **1983**, *72*, 691–707. [CrossRef]
3. Zhao, Z. Internal tide radiation from the Luzon Strait. *J. Geophys. Res.* **2014**, *119*, 5434–5448. [CrossRef]
4. Helfrich, K.R.; Grimshaw, R.H.J. Nonlinear disintegration of the internal tide. *J. Phys. Oceanogr.* **2008**, *38*, 686–701. [CrossRef]
5. Jia, T.; Liang, J.J.; Li, X.-M.; Sha, J. SAR observation and numerical simulation of internal solitary wave refraction and reconnection behind the Dongsha Atoll. *J. Geophys. Res. Oceans* **2018**, *123*, 74–89. [CrossRef]
6. Chang, M.-H.; Lien, R.-C.; Tang, T.Y.; D'Asaro, E.A.; Yang, Y.J. Energy flux of nonlinear internal waves in northern South China Sea. *Geophys. Res. Lett.* **2006**, *33*. [CrossRef]
7. Lien, R.-C.; D'Asaro, E.A.; Henyey, F.; Chang, M.-H.; Tang, T.-Y.; Yang, Y.-J. Trapped Core Formation within a Shoaling Nonlinear Internal Wave. *J. Phys. Oceanogr.* **2012**, *42*, 511–525. [CrossRef]
8. Liu, A.K.; Chang, Y.S.; Hsu, M.K.; Liang, N.K. Evolution of nonlinear internal waves in the East and South China Seas. *J. Geophys. Res.* **1998**, *103*, 7995–8008. [CrossRef]
9. Liu, A.K.; Hsu, M.K. Internal wave study in the South China Sea using Synthetic Aperture Radar (SAR). *Int. J. Remote Sens.* **2004**, *25*, 1261–1264. [CrossRef]
10. Alpers, W.; He, M.-X.; Zeng, K.; Guo, L.-F.; Li, X.-M. The distribution of internal waves in the China Seas studied by multi-sensor satellite images. In Proceedings of the 2005 Dragon Symposium, Santorini, Greece, 27 June 27–1 July 2005.
11. Xu, Z.H.; Yin, B.S.; Hou, Y.J. Highly nonlinear internal solitary waves over the continental shelf of the northwestern South China Sea. *Chin. J. Ocean. Limnol.* **2010**, *28*, 1049–1054. [CrossRef]
12. Alford, M.H.; Mickett, J.B.; Zhang, S.; MacCready, P.; Zhao, Z.; Newton, J. Internal waves on the Washington continental shelf. *Oceanography* **2012**, *25*, 66–79. [CrossRef]
13. Zhao, Z.; Klemas, V.; Zheng, Q.; Yan, X.-H. Remote sensing evidence for baroclinic tide origin of internal solitary waves in the northeastern South China Sea. *Geophys. Res. Lett.* **2004**, *31*. [CrossRef]
14. Li, X.F.; Zhao, Z.X.; Pichel, W.G. Internal solitary waves in the northwestern South China Sea inferred from satellite images. *Geophys. Res. Lett.* **2008**, *35*. [CrossRef]
15. Li, Q.; Farmer, D.M. The generation and evolution of nonlinear internal waves in the deep basin of the South China Sea. *J. Phys. Oceanogr.* **2011**, *41*, 1345–1363. [CrossRef]
16. Xu, Z.H.; Yin, B.S.; Hou, Y.J.; Fan, Z.S.; Liu, A.K. A study of internal solitary waves observed on the continental shelf in the northwestern South China Sea. *Acta Oceanol. Sin.* **2010**, *29*, 18–25. [CrossRef]
17. Whitham, G.B. *Linear and Nonlinear Waves*; John Wiley & Sons, Inc.: Hoboken, NJ, USA, 1999.
18. Jackson, C.R. An empirical model for estimating the geographic location of nonlinear internal solitary waves. *J. Atmos. Ocean. Technol.* **2009**, *26*, 2243–2255. [CrossRef]
19. Alford, M.H.; MacKinnon, J.A.; Nash, J.D.; Simmons, H.; Pickering, A.; Klymak, J.M.; Pinkel, R.; Sun, O.; Rainville, L.; Musgrave, R.; et al. Energy flux and dissipation in Luzon strait: Two tales of two ridges. *J. Phys. Oceanogr.* **2011**, *41*, 2211–2222. [CrossRef]
20. Sethian, J.A. *Level Set Methods and Fast Marching Methods*; Cambridge University Press: Cambridge, UK, 1999.
21. Egbert, G.D.; Erofeeva, S.Y. Efficient Inverse Modeling of Barotropic Ocean Tides. *J. Atmos. Ocean. Technol.* **2002**, *19*, 183–204. [CrossRef]
22. Xie, X.-H.; Chen, G.-Y.; Shang, X.-D.; Fang, W.-D. Evolution of the semidiurnal (M2) internal tide on the continental slope of the northern South China Sea. *Geophys. Res. Lett.* **2008**, *35*. [CrossRef]
23. Liu, Q.; Shang, X.; Xie, X. Observations of semidiurnal M2 internal tidal parametric subharmonic instability in the northeastern South China Sea. *J. Ocean. Limnol.* **2021**, *39*, 56–63. [CrossRef]
24. Li, X.F.; Jackson, C.R.; Pichel, W.G. Internal solitary wave refraction at Dongsha Atoll, South China Sea. *Geophys. Res. Lett.* **2013**, *40*, 3128–3132. [CrossRef]

Journal of
Marine Science and Engineering

MDPI

Article

A Strong Internal Solitary Wave with Extreme Velocity Captured Northeast of Dong-Sha Atoll in the Northern South China Sea

Andi Xu and Xueen Chen *

College of Oceanic and Atmospheric Sciences, Ocean University of China, Qingdao 266100, China; xad@stu.ouc.edu.cn
* Correspondence: xchen@ouc.edu.cn

Abstract: Internal solitary waves (ISWs) in the South China Sea (SCS) have received considerable attention. This paper reports on a strong ISW captured northeast of Dong-Sha Atoll on 22 May 2011 by shipboard Acoustic Doppler Current Profiler (ADCP), which had the largest velocity among the ISWs so far reported in the global ocean. The peak westward velocity (u) was 2.94 m/s, and the peak downward velocity (w) was 0.63 m/s, indicating a first baroclinic mode depression wave. The amplitude of ISW inferred from ADCP backscatter was about 97 m. 2.2 h later, a trailing wave was captured with a peak westward velocity and downward velocity of 2.24 m/s and 0.42 m/s, respectively, surprisingly large for a trailing wave, suggesting that the ISW is type-A wave. The estimated baroclinic current induced by the leading ISW was much larger than the barotropic current. The Korteweg-De Vries (KdV) theoretical phase speed and the phase speed inferred from the satellite images were 1.76 m/s and 1.59 m/s, respectively. The peak horizontal velocity exceeded the phase speed, suggesting the ISW was close to or already in the process of breaking and may have formed a trapped core.

Keywords: internal solitary wave; South China Sea; shipboard observation; extreme current velocity; wave breaking; trapped core

Citation: Xu, A.; Chen, X. A Strong Internal Solitary Wave with Extreme Velocity Captured Northeast of Dong-Sha Atoll in the Northern South China Sea. *J. Mar. Sci. Eng.* **2021**, *9*, 1277. https://doi.org/10.3390/jmse9111277

Academic Editor: Lev Shemer

Received: 7 October 2021
Accepted: 10 November 2021
Published: 17 November 2021

1. Introduction

Internal solitary waves (ISWs) are ubiquitous in some marginal seas, especially in the South China Sea (SCS) [1]. They are supposedly generated near the Batan Islands of Luzon Strait and propagate northwestward in the deep basin with long crest lengths ranging from about 100 to 250 km [2,3], then shoal onto the continental shelf and finally dissipate on the plateau [4].

ISWs have rapid spatial evolution in shallow water [5], and they have been extensively studied near Dong-Sha Atoll by in situ measurements, remote sensing, and numerical simulations. Strong ISWs have been observed near Dong-Sha Atoll by moorings and shipboard surveys with amplitudes ranging from 70 to 173 m and peak westward velocities from 0.73 to 2.4 m/s [4–12].

However, one ISW with a peak westward current velocity of 2.94 m/s northeast of Dong-Sha Atoll is presented in this paper, with the largest ISW velocity ever observed in the global ocean.

This paper is organized as follows. Section 2 gives a general representation of the data. The description of measurements and the phase speed analysis are provided in Section 3. The discussion follows in Section 4, which includes the impact of the barotropic tide, the trapped core, and the breaking stage. Finally, a conclusion is given in Section 5.

2. Materials

2.1. Field Measurements

The shipboard observations were obtained near the continental shelf break to the northeast of Dong-Sha Atoll, during 21–22 May 2011 (Figure 1). The research vessel collected temperature, conductivity, and depth (CTD) measurements with a Sea-Bird SBE-911 and velocity, echo amplitude with a hull-mounted 75 kHz Teledyne RDI Acoustic Doppler Current Profiler (ADCP). The ADCP vertical resolution was 16 m over a depth range from 25 m to about 650 m. The temporal resolution was 2 s, recording ensemble averages every minute, suitable to sample the ISW with the period in the order of 10 min. Several ISWs were captured during this cruise. This paper focuses on an ISW with an extreme velocity captured on 22 May that was selected for extensive discussion.

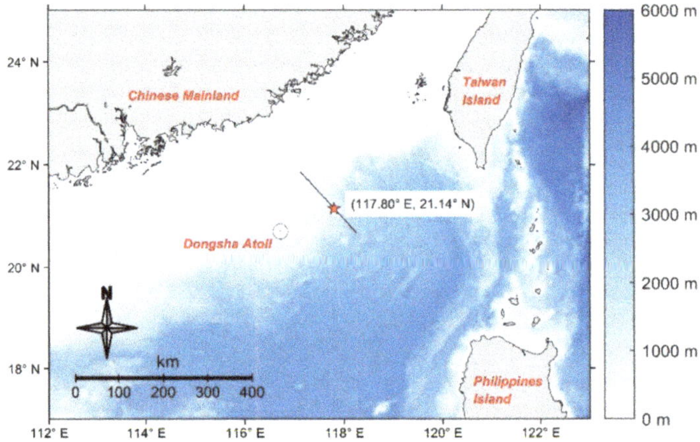

Figure 1. The topography in the South China Sea. The vessel's track between 18:00 UTC on 21 May and 08:00 UTC on 22 May is located within the black line area. The red pentagram gives the location where the internal solitary wave (ISW) was captured at 00:00 UTC on 22 May 2011.

2.2. HYCOM Reanalysis Products

The background stratification was derived from a HYbrid Coordinate Ocean Model (HYCOM) of temperature and salinity with a temporal resolution of 3 h and a spatial resolution of 0.08°. The variables are interpolated onto 41 layers at depths between 0 and 5000 m. The model assimilates altimeter observations and in situ sea surface temperature, as well as vertical temperature and salinity profiles from Argo floats, expendable bathythermographs (XBTs), and moored buoys, by Navy Coupled Ocean Data Assimilation (NCODA) system [13,14]. The model data was used to estimate stratification before the ISW event.

2.3. Satellite Data

Remote sensing data such as Synthetic Aperture Radar (SAR) and Advanced Synthetic Aperture Radar (ASAR) has been widely used to study ISWs [15]. Depression ISWs can be identified on SAR images as a bright band in front followed by a dark band on SAR images, and vice-versa for elevation ISWs [16].

European Space Agency Environment satellite (Envisat) images were used to investigate sea surface signatures of ISWs near Dong-Sha Atoll in the northern SCS. The Envisat is an advanced polar-orbiting Earth observation satellite with ASAR active between 1 March 2002 and 8 April 2012, with a spatial resolution of about 30 m, sufficient to resolve the ISWs.

3. Results

3.1. Vertical Profiles of Velocity and Echo Amplitude

On 22 May 2011, an unusual ISW with extreme velocity was captured between 00:00 and 00:45 UTC in northern SCS to the northeast of the Dong-Sha Atoll, where the depth is 880 m (shown in Figure 1). Figure 2 shows the research vessel's track and the navigation speed from 23:48 UTC on 21 May to 03:00 UTC on 22 May. During the above time slot, the vessel drifted from (117.795° E, 21.146° N) to (117.753° E, 21.154° N) when the ISW passed by (red line in Figure 2a). Then the vessel moved back to the primary site near (117.795° E, 21.146° N) (black line in Figure 2a) and captured a trailing wave about 2.2 h later than the leading wave. Before the trailing wave arrived, the vessel was drifting (blue line in Figure 2a). The ADCP was switched off between 00:48 and 01:26 UTC, so the ship's track and speed in this period are not shown. The averaged drifting speed and the maximum drifting speed induced by the leading wave were 1.62 m/s and 2.53 m/s, respectively. For the trailing wave, the corresponding speeds were 1.51 m/s and 1.96 m/s, which were uncommonly large.

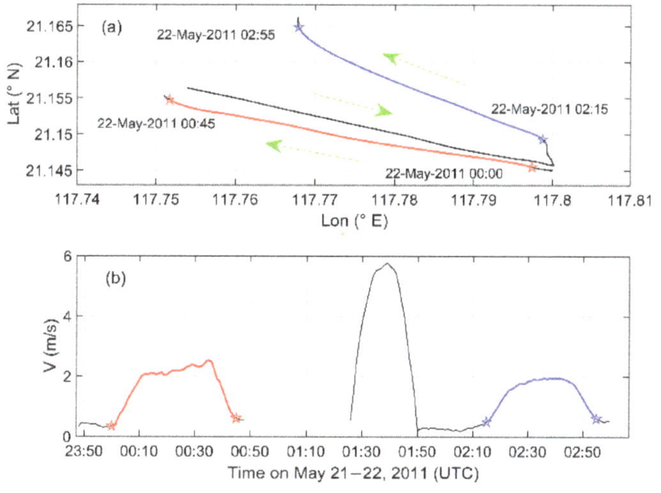

Figure 2. (**a**) Vessel's track; (**b**) vessel's navigation speed. The red and the blue lines correspond to the time of the leading wave and the trailing wave, respectively. The pentagrams indicate the beginning and the ending of the time of the waves. The green arrows indicate the vessel's direction. The blank from 00:48 UTC to 01:26 UTC corresponds to the ADCP being switched off.

The mean ADCP velocity data 30 min prior to the ISW arrival is used as the background current, not influenced by the ISW yet. The velocity profile of the ISW was calculated by subtracting the background current. Figure 3 shows current velocity and echo amplitude timeseries, as a function of depth from 23:17 to 00:48 UTC. Notable horizontal current velocities were observed near the surface, with a peak westward velocity (u) of 2.94 m/s, where the depth was 89 m (Figure 3a). There were strong downward (upward) currents (w) at the leading (trailing) edge, and the peak velocity was 0.63 m/s (Figure 3b) extended in the whole water column. The northward velocity (v) was small and not the focus of this paper. One instance of significant horizontal velocity shear was recorded (Figure 3a), suggesting this ISW was a first baroclinic mode depression wave [2].

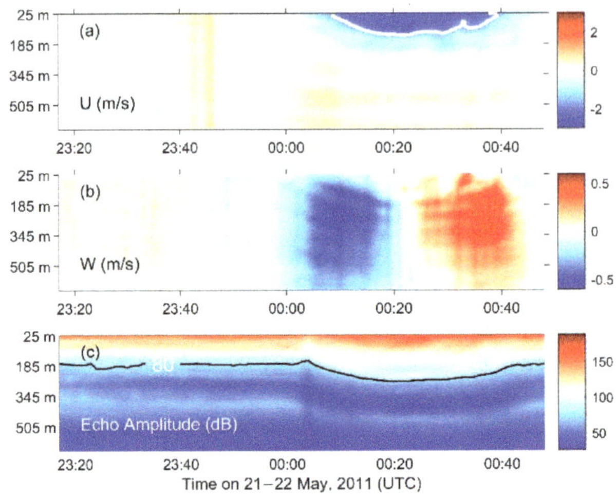

Figure 3. Current velocity and echo amplitude profiles of the ISW. (**a**) Zonal current velocity. The thick white line indicates the phase speed of ISW estimated from Korteweg-De Vries (KdV) theory. (**b**) Vertical current velocity. (**c**) Echo amplitude; the 80 dB isopycnal is highlighted by the black lines.

The echo amplitude from ADCP closely tracks isopycnals inferred from seawater density and biomass on a large scale, which can be qualitatively used for visualizing the internal waves [4,17]. The maximum vertical excursion of echo amplitude reached 97 m, located at a depth of 185 m, suggesting the amplitude of ISW was 97 m (Figure 3c).

Figure 4 shows the trailing wave current velocity timeseries. The signal around 01:50 UTC was induced by a sudden deceleration of the ship (Figure 2b). The trailing wave had a peak westward velocity of 2.24 m/s at a depth of 73 m and peak downward velocity of 0.42 m/s; surprisingly large for a trailing wave.

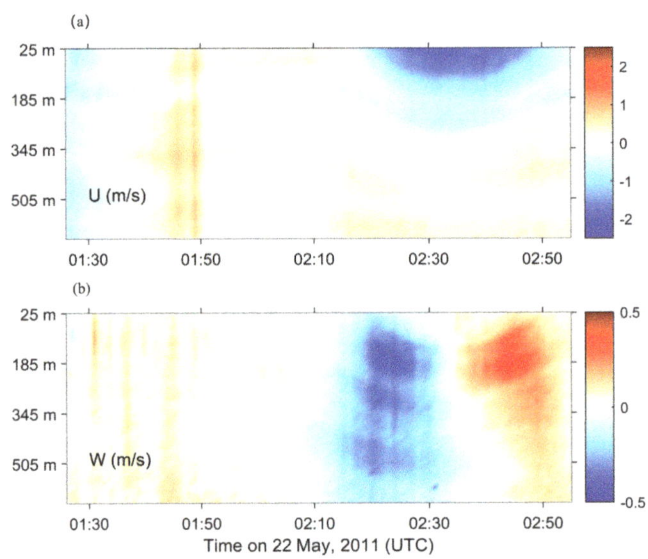

Figure 4. Current velocity profiles of the trailing wave. (**a**) Zonal current velocity; (**b**) vertical current velocity.

3.2. Analysis of Phase Speed

3.2.1. Solution of the Korteweg-De Vries (KdV) Theory

Since no CTD cast was deployed near the location of ISW, the background stratification was calculated from HYCOM data at 00:00 UTC (Figure 5). The temperature profile shows a main thermocline between 35 and 300 m, below a mixed layer with a depth-independent temperature of 26.2 °C. The background current shows there was a slightly westward flow with a mean speed of 0.06 m/s. The calculated Brunt–Väisälä frequency N^2 indicates that the strongest stratification was at a depth of 37.5 m.

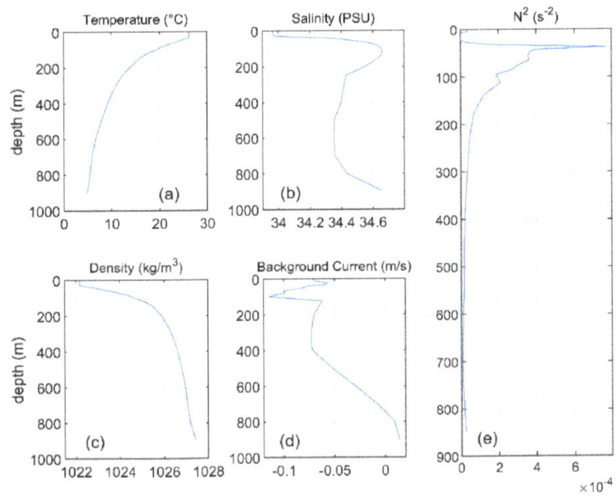

Figure 5. The background physical properties of the HYbrid Coordinate Ocean Model (HYCOM) product. (**a–e**) Temperature, salinity, density calculated, background current, and Brunt–Väisälä frequency N^2.

The wave is fitted to the KdV equation in a stratified fluid, which is generally used to describe characteristics of ISW [18].

$$\frac{\partial \eta}{\partial t} + c_0\left(\frac{\partial \eta}{\partial x} + \alpha\eta\frac{\partial \eta}{\partial x} + \beta\frac{\partial^3 \eta}{\partial x^3}\right) = 0 \tag{1}$$

In Equation (1), where c_0 is the linear wave speed and η is the vertical displacement of the ISW. The parameters α and β are the nonlinear parameter and dissipation parameter, respectively. These two coefficients are also called "environmental parameters" as they contribute to conditions such as stratification and water depth [19].

The environmental parameters can be calculated by as the Equations (2) and (3).

$$\alpha = \frac{3}{2}\frac{\int_{-H}^{0}\rho_0(z)\left(\frac{df_n}{dz}\right)^3 dz}{\int_{-H}^{0}\rho_0(z)\left(\frac{df_n}{dz}\right)^2 dz} \tag{2}$$

$$\beta = \frac{1}{2}\frac{\int_{-H}^{0}\rho_0(z)(f_n)^2 dz}{\int_{-H}^{0}\rho_0(z)\left(\frac{df_n}{dz}\right)^2 dz} \tag{3}$$

where ρ_0 is the density of the depth z, and where f_n is the vertical structure of vertical displacement corresponding to a particular mode n, which can be solved by the boundary condition problem [20].

$$\frac{d}{dz}[\rho_0(z)\frac{df_n}{dz}] + \rho_0(z)\frac{N^2(z)}{c_n^2}f_n = 0 \tag{4}$$

$$f_n(z) = 0,\ z = 0 \tag{5}$$

$$f_n(z) = 0,\ z = -H \tag{6}$$

where N^2 is the Brunt–Väisälä frequency, where H is the bottom depth.

The solution to the KdV equation is

$$\eta(x,t) = \eta_0 sech^2\left(\frac{x - vt}{\Delta}\right) \tag{7}$$

where η_0 is the wave amplitude, where $\Delta = \sqrt{\frac{12\beta}{\alpha\eta_0}}$ is the characteristic width, and the phase speed $c_p = c_0\left(1 + \frac{1}{3}\alpha\eta_0\right)$.

According to the formulas above, the calculated nonlinear parameter α is approximately $-5.42 \times 10^{-4}\ s^{-1}$, and the calculated dissipation parameter β is $4.74 \times 10^4\ m^3\ s^{-1}$. The negative sign of the nonlinear parameter α suggests it is a depression wave [21]. The calculated characteristic width of the wave Δ is 3289.4 m, and the calculated phase speed c_p is 1.76 m/s.

3.2.2. Estimation by Satellite Image

Fortunately, we got one ASAR image during the cruise. The phase speed was also estimated from the satellite image for comparison. Figure 6 shows an ASAR image acquired by the Envisat satellite at 02:17 UTC on 23 May 2011, capturing the same ISW 26.3 h after it was detected by the vessel. The image shows several bright–dark bands corresponding to wave crests–troughs near Dong-sha Atoll, indicating the first baroclinic mode depression waves. The wave we discussed above was a leading wave, and the trailing wave was also detected (Section 3). The ISW wave packet can be classified as an A-type wave.

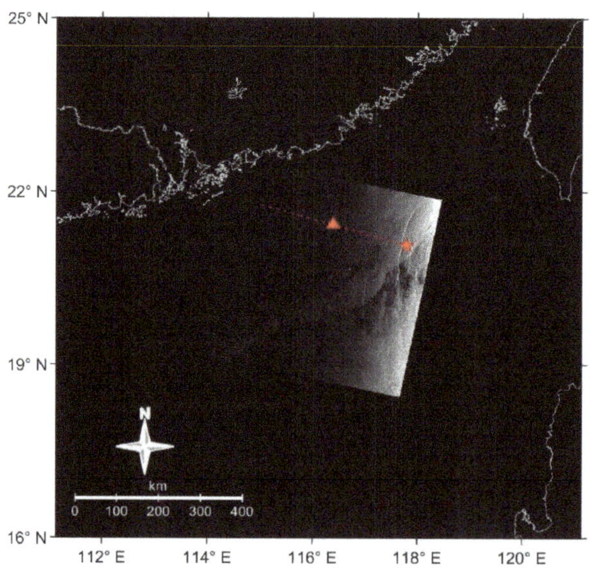

Figure 6. Satellite image of the Environment satellite (Envisat) taken at 02:17 UTC on 23 May 2011, 26.3 h after the ISW was measured by the shipboard ADCP. The red pentagram represents the location where the ISW was captured by the ship, the dashed line represents the propagation path, and the red triangle represents the same ISW 26.3 h after it was detected in the satellite image.

The distance between the wave crest (shown in Figure 6) and the ship ADCP measurements location was about 150 km. Considering the distance and time difference, the mean phase speed of the ISW was around 1.59 m/s, which is only 0.17 m/s slower than the theoretical result of the KdV solution. The difference can be attributed to the accuracy of HYCOM data products in the SCS, and also may relate to the variations of bathymetry over a distance of 150 km and the background current.

4. Discussion

4.1. Influence of Barotropic Tide

The barotropic tide velocity can reach in the order of 10 cm/s in the northern SCS [22], which may contribute to the horizontal velocity in ADCP measurements. When considering the influence of the barotropic tide on the observed ISW, the barotropic tide should be calculated first. With long-term and all-depth observation records, the internal wave signal could be separated via filter analysis from tidal and inertial signals [2,23,24] to investigate the influence of the barotropic tide on the observed ISW. However, the vessel cannot provide full-depth current data due to ADCP measurement limitations. Instead, here we considered two other methods to subtract the velocity of the barotropic tide.

4.1.1. Modal Decomposition to Reconstruct the Barotropic Current

According to [25], full-depth barotropic currents and baroclinic currents from limited mooring observations can be obtained by combining harmonic analysis and modal decomposition. Harmonic analysis is used to calculate harmonic constants of the major constituents and predict the time series of each major tidal constituent. Modal decomposition is carried out to obtain full-depth tidal currents of each mode by using the least-squares method [26]. It was found that when more than three vertical modes were considered in the calculation, the prescribed tidal currents could be reconstructed accurately [25].

As the time series of shipboard ADCP measurement was not long enough to be frequency filtered, processes at frequencies other than barotropic tide and baroclinic tide were ignored. Thus, we focused on the modal decomposition, only taking into account the first three vertical modes. The calculation did not go into detail here.

The first three vertical modes were taken into account. The reconstructed barotropic westward velocity (u) was 2 cm/s, and the southward velocity (v) was 6.34 cm/s. Both were insignificant compared to the large velocity induced by ISW.

4.1.2. Estimation of Barotropic Current Using Numerical Modelling

Numerical modeling can also be used to infer the tidal signal [27]. Here we use the Tidal Model Driver (TMD) toolbox to predict the barotropic tidal constituent height and amplitude [28].

At 00:00 UTC on 22 May, the prediction of westward barotropic tidal velocity (u) was 6.36 cm/s, and the southward velocity (v) was 3.12 cm/s, which are similar to the results mentioned above.

In conclusion, the barotropic tidal velocity at the site where the ISW was observed likely has little influence on the current field.

4.2. The Trapped Core and Breaking Stage

The trapped core is defined as the particle, which was trapped in the wave and carried along at constant propagating speed [29] and occurs when the water velocity exceeds the propagation speed of the wave [30]. It is formed when ISWs shoal from deep to shallow water. As the ISW reaches its breaking limit, it forms a subsurface trapped core [31]. A trapped core was identified in the measured ISW zonal velocity (inside the white line representing the ISW phase speed in Figure 3a). The upper portion of the trapped core was beyond the shipboard ADCP range.

The breaking stage can be defined according to [32] by comparing the peak current velocity u_{max} and phase speed c. When $u_{max} > 0.8\,c$ in a continuously stratified fluid, the ISW would break. The ISW presented in this paper had values of $u_{max} = 2.93$ m/s and $c = 1.76$ m/s, with peak current velocity much larger. Obviously, the peak current velocity was much larger than the estimated phase speed. The ISW was close to or already in the process of breaking.

The above results are consistent with those in [5], which present 41 ISWs captured in April 1999 and 2000 on the continental slope of the northern South China Sea during the Asian Seas International Acoustics Experiment (ASIAEX) [5]. One ISW among them was similar to our observations, with the largest westward current exceeding 2.4 m/s right near our research area (details shown in Table 1). This suggests that when one ISW approaches the breaking stage, its velocity could be significantly larger than before.

Table 1. Observations in this study and in [5].

	Location	Depth	Peak Western Current Velocity	Depth of Nodal Point	Breaking Stage
This study	(117.80° E, 21.14° N)	880 m	22 May 2011	2.94 m/s	Between 297 m and 329 m
Results in [5]	(117.22° E, 21.05° N)	426 m	9 April 2000	Missing data, but exceeding 2.4 m/s	Between 210 m and 323 m

5. Conclusions

In this paper, an ISW with extreme current velocity northeast of Dong-Sha Atoll is reported and discussed, whose velocity is the largest among the ISWs ever observed in the global ocean. The peak westward velocity (u) was 2.94 m/s, and the peak downward velocity (w) was 0.63 m/s, respectively. The amplitude of ISW was about 97 m, propagating as a mode 1 depression wave. A strong trailing wave was captured 2.2 h later and suggested the ISW was an A-type wave. The barotropic tide had little influence on the current field of the ISW, which indicated that the velocity induced by the ISW was extremely large indeed. The KdV theoretical phase speed and the estimated phase speed by the satellite image were 1.76 m/s and 1.59 m/s, respectively. By comparing the peak current velocity u_{max} and phase speed c, it is inferred that the ISW reported in our paper was close to or already in the process of breaking and may have developed a trapped core.

In the deep basin of northern SCS, a strong ISW was reported with an amplitude of 240 m and a peak westward current velocity of 2.55 m/s [33], with a larger amplitude but lower peak velocity than the one discussed in this paper. However, the ISW reported in [33] was captured in the deep basin of northern SCS where the depth was 3847 m, rather than in the vicinity of Dong-sha Atoll.

ISWs with such a large velocity are rare, with previous observations near Dong-Sha Atoll reporting amplitudes ranging from 70 to 106 m and peak westward velocities from 0.73 to 2.4 m/s. In the vicinity of Dong-sha Atoll, 41 ISWs were identified with moorings, among which only one ISW with a large peak westward current exceeding 2.4 m/s was captured, whose characteristics were similar to the ISW discussed in this paper [5]. Both ISWs in the two studies were near breaking. Thus, the ISW discussed in this paper has the largest velocity so far reported in the global ocean.

According to [11], ISWs in a marginal convectively unstable state (which leads to a trapped core) can contribute to turbulence mixing with long distances, along with the propagation. We supposed that the turbulence mixing would appear if the breaking state continued, but further observations are needed to prove it.

Author Contributions: Conceptualization, A.X. and X.C.; data curation, A.X.; formal analysis, A.X.; investigation, A.X.; methodology, A.X.; software, A.X.; validation, A.X.; visualization, A.X.; funding acquisition, X.C.; resources, X.C.; supervision, X.C.; writing—original draft, A.X.; writing—review and editing, A.X. and X.C. All authors have read and agreed to the published version of the manuscript.

Funding: This research was funded by the National Key Research and Development Plan, Grant 2016YFC1401300, "Oceanic Instruments Standardization Sea Trials (OISST)", and the Taishan Scholar Program.

Institutional Review Board Statement: Not applicable.

Informed Consent Statement: Not applicable.

Data Availability Statement: The in situ data presented in this study are available on request from the corresponding author.

Acknowledgments: The authors appreciate the National Ocean Partnership Program (NOPP) for providing the HYCOM reanalysis products and the Open spatial data sharing system of the Aerospace Information Research Institute (AIR) for providing the satellite image data. We are grateful to the Earth & Space Research (ESR) and Oregon State University (OSU) for providing the Tidal Model Driver (TMD) Matlab toolbox. We also thank the helpful suggestions from two anonymous reviewers for improving the manuscript.

Conflicts of Interest: The authors declare no conflict of interest.

References

1. Jackson, C. Internal wave detection using the Moderate Resolution Imaging Spectroradiometer (MODIS). *J. Geophys. Res.* **2007**, *112*. [CrossRef]
2. Ramp, S.R.; Tang, T.Y.; Duda, T.F.; Lynch, J.F.; Liu, A.K.; Chiu, C.S.; Bahr, F.L.; Kim, H.R.; Yang, Y.J. Internal Solitons in the Northeastern South China Sea Part I: Sources and Deep Water Propagation. *IEEE J. Ocean. Eng.* **2004**, *29*, 1157–1181. [CrossRef]
3. Yu, C.; Feng, Z.; Jun, Z. Numerical simulation of internal waves excited by a submarine moving in the two-layer stratified fluid. *J. Hydrodyn.* **2006**, *18*, 330–336. [CrossRef]
4. Klymak, J.M.; Pinkel, R.; Liu, C.T.; Liu, A.K.; David, L. Prototypical solitons in the South China Sea. *Geophys. Res. Lett.* **2006**, *33*. [CrossRef]
5. Yang, Y.J.; Tang, T.Y.; Chang, M.H.; Liu, A.K.; Hsu, M.K.; Ramp, S.R. Solitons Northeast of Tung-Sha Island During the ASIAEX Pilot Studies. *IEEE J. Ocean. Eng.* **2004**, *29*, 1182–1199. [CrossRef]
6. Fang, W.; Ping, S.; Xiaomin, L.; Qingwen, M. In situ observations of internal solitary waves in the northern South China Sea. *Chin. Sci. Bull.* **2005**, *50*, 1400–1404. [CrossRef]
7. Zhao, W.; Huang, X.; Tian, J. A new method to estimate phase speed and vertical velocity of internal solitary waves in the South China Sea. *J. Oceanogr.* **2012**, *68*, 761–769. [CrossRef]
8. Lien, R.; Henyey, F.; Ma, B.; Yang, Y.J. Large-Amplitude Internal Solitary Waves Observed in the Northern South China Sea: Properties and Energetics. *J. Phys. Oceanogr.* **2014**, *44*, 1095–1115. [CrossRef]
9. Chen, L.; Zheng, Q.; Xiong, X.; Yuan, Y.; Xie, H.; Guo, Y.; Yu, L.; Yun, S. Dynamic and Statistical Features of Internal Solitary Waves on the Continental Slope in the Northern South China Sea Derived from Mooring Observations. *J. Geophys. Res. Oceans* **2019**, *124*, 4078–4097. [CrossRef]
10. Chang, M.; Cheng, Y.; Yang, Y.J.; Jan, S.; Ramp, S.R.; Reeder, D.B.; Hsieh, W.; Ko, D.S.; Shao, H.; Tseng, R. Instabilities and Turbulence Observed within Large Internal Solitary Waves in the South China Sea. In Proceedings of the Ocean Sciences Meeting 2020 (AGU 2020), San Diego, CA, USA, 16–21 February 2020.
11. Chang, M.H.; Lien, R.C.; Lamb, K.G.; Diamessis, P.J. Long-Term Observations of Shoaling Internal Solitary Waves in the Northern South China Sea. *J. Geophys. Res. Oceans* **2021**, *126*, e2020JC017129. [CrossRef]
12. Kuang, Y.; Wang, Y.; Song, H.; Guan, Y.; Fan, W.; Yi, G.; Zhang, K. Study of internal solitary wave packets in the northeastern South China Sea based on seismic oceanography and remote sensing. *Chin. J. Geophys. Chin. Ed.* **2021**, *64*, 597–611. [CrossRef]
13. Cummings, J.A. Operational multivariate ocean data assimilation. *Q. J. R. Meteorol. Soc.* **2005**, *131*, 3583–3604. [CrossRef]
14. Cummings, J.A.; Smedstad, O.M. Variational Data Assimilation for the Global Ocean. In *Data Assimilation for Atmospheric, Oceanic and Hydrologic Applications (Vol. II)*; Park, S.K., Xu, L., Eds.; Springer: Berlin/Heidelberg, Germany, 2013; pp. 303–343.
15. Klemas, V. Remote Sensing of Coastal Plumes and Ocean Fronts: Overview and Case Study. *J. Coast. Res.* **2012**, *278*, 1–7. [CrossRef]
16. Liu, A.K. Analysis of nonlinear internal waves in the New York Bight. *J. Geophys. Res. Oceans* **1988**, *93*, 12317–12329. [CrossRef]
17. Dowd, M.; Thompson, K.R. Extraction of tidal streams from a ship-borne acoustic Doppler current profiler using a statistical-dynamical model. *J. Geophys. Res. Oceans* **1996**, *101*, 8943–8956. [CrossRef]
18. Ostrovsky, L.A.; Stepanyants, Y.A. Do internal solitons exist in the ocean? *Rev. Geophys.* **1989**, *27*, 293–310. [CrossRef]
19. Lee, C.; Beardsley, R.C. The generation of long nonlinear internal waves in a weakly stratified shear flow. *J. Geophys. Res.* **1974**, *79*, 453–462. [CrossRef]
20. Apel, J.R.; Ostrovsky, L.A.; Stepanyants, Y.A.; Lynch, J.F. Internal solitons in the ocean and their effect on underwater sound. *J. Acoust. Soc. Am.* **2007**, *121*, 695–722. [CrossRef]
21. Helfrich, K.R.; Melville, W.K. On long nonlinear internal waves over slope-shelf topography. *J. Fluid Mech.* **1986**, *167*, 285–308. [CrossRef]
22. Yang, W.; Yin, B.; Yang, D.; Xu, Z. Application of FVCOM in numerical simulation of tide and tidal currents in the northern South China. *Mar. Sci.* **2013**, *37*, 10–19.

23. Xu, Z.; Yin, B.; Hou, Y.; Fan, Z.; Liu, A.K. A study of internal solitary waves observed on the continental shelf in the northwestern South China Sea. *Acta Oceanol. Sin.* **2010**, *29*, 18–25. [CrossRef]
24. Qian, H.; Huang, X.; Tian, J. Observational study of one prototypical mode-2 internal solitary waves in the northern South China Sea. *Acta Oceanol. Sin.* **2016**, *38*, 13–20. [CrossRef]
25. Cao, A.; Li, B.; Lv, X. Extraction of Internal Tidal Currents and Reconstruction of Full-Depth Tidal Currents from Mooring Observations. *J. Atmos. Ocean. Technol.* **2015**, *32*, 1414–1424. [CrossRef]
26. Griffiths, S.D.; Grimshaw, R.H.J. Internal tide generation at the continental shelf modeled using a modal decomposition: Two-dimensional results. *J. Phys. Oceanogr.* **2007**, *37*, 428–451. [CrossRef]
27. Howarth, M.J.; Proctor, R. Ship ADCP measurements and tidal models of the North Sea. *Cont. Shelf Res.* **1992**, *12*, 601–623. [CrossRef]
28. Earth and Space Research/TMD_Matlab_Toolbox_v2.5. Available online: https://github.com/EarthAndSpaceResearch/TMD_Matlab_Toolbox_v2.5 (accessed on 3 October 2021).
29. Lamb, K.G. A numerical investigation of solitary internal waves with trapped cores formed via shoaling. *J. Fluid Mech.* **2002**, *451*, 109–144. [CrossRef]
30. Lien, R.; Asaro, E.A.D.; Henyey, F.; Chang, M.; Tang, T.; Yang, Y. Trapped Core Formation within a Shoaling Nonlinear Internal Wave. *J. Phys. Oceanogr.* **2012**, *42*, 511–525. [CrossRef]
31. Lamb, K.G. Shoaling solitary internal waves: On a criterion for the formation of waves with trapped cores. *J. Fluid Mech.* **2003**, *478*, 81–100. [CrossRef]
32. Grue, J.; Jensen, A.; Rusas, P.; Sveen, J.K. Breaking and broadening of internal solitary waves. *J. Fluid Mech.* **2000**, *413*, 181–217. [CrossRef]
33. Huang, X.; Chen, Z.; Zhao, W.; Zhang, Z.; Zhou, C.; Yang, Q.; Tian, J. An extreme internal solitary wave event observed in the northern South China Sea. *Sci. Rep.* **2016**, *6*, 30041. [CrossRef]

Journal of
Marine Science and Engineering

MDPI

Article

Noise of Internal Solitary Waves Measured by Mooring-Mounted Hydrophone Array in the South China Sea

Jiemeihui Li [1,2], Yang Shi [1,2,*], Yixin Yang [1,2] and Xiaodong Huang [3]

1 School of Marine Science and Technology, Northwestern Polytechnical University, Xi'an 710000, China; ljmh1994@mail.nwpu.edu.cn (J.L.); yxyang@nwpu.edu.cn (Y.Y.)
2 Shaanxi Key Laboratory of Underwater Information Technology, Northwestern Polytechnical University, Xi'an 710000, China
3 Physical Oceanography Laboratory, Ocean University of China, Qingdao 266000, China; xhuang@ouc.edu.cn
* Correspondence: shiyang@nwpu.edu.cn

Abstract: Internal solitary waves in the South China Sea have attracted attention because of their large amplitude and high rate of occurrence. Internal solitary waves have a substantial influence on underwater sound propagation and ambient noise. However, there have seldom been reports on the noise they cause. In this paper, we conducted an internal solitary waves cooperative observation experiment in the South China Sea in 2016. We analyzed the temperature, flow velocity and noise changes induced by internal solitary waves. The power spectra of noise generated by internal solitary waves at frequencies below 100 Hz was almost 20 dB higher than ambient noise. The observed low-frequency noise had uniform harmonics. Combined with the changes of flow velocity, we interpreted the low frequency noise as flow noise induced by vortex-induced vibration of internal solitary waves flowing past the cable mooring system. The noise spectra were related to the position of the cable where the hydrophone was mounted. The closer they were to the middle of the cable, the greater the vibration amplitude, and the stronger the noise. This study provided a passive acoustic monitoring and warning method for high marine currents.

Keywords: internal solitary waves; underwater noise; flow noise; vortex-induced vibration; the South China Sea

Citation: Li, J.; Shi, Y.; Yang, Y.; Huang, X. Noise of Internal Solitary Waves Measured by Mooring-Mounted Hydrophone Array in the South China Sea. *J. Mar. Sci. Eng.* **2022**, *10*, 222. https://doi.org/10.3390/jmse10020222

Academic Editor: Unai Fernandez-Gamiz

Received: 22 December 2021
Accepted: 4 February 2022
Published: 8 February 2022

1. Introduction

Internal solitary waves (ISWs), characterized by their large amplitudes, high rate of occurrence and strong nonlinearity, are widely distributed in global oceans with crest lengths of up to 200 km [1]. As a result of tide-topography interactions, ISWs are especially active in marginal seas and around straits where strong tidal currents flow over steep topographies [2,3]. In those areas, ISWs induce horizontal current velocities exceeding 2 m/s, and in vertical currents, ISWs depresse isopycnal surfaces rapidly with fluctuating amplitudes of up to 240 m in 10 min [4–7]. Moreover, ISWs in the oceans often appear as multi-wave packets which contain a number of rank-ordered solitary waves in groups [3,7].

The South China Sea is a marginal sea with a high concentration and occurrence of ISWs [8]. ISWs in the northern South China Sea are among the strongest waves in global oceans. As such, recent years have seen growing attention paid to ISWs in the South China Sea. Xu and Chen reported a strong ISW in the northeast of Dong-sha Atoll, which had a velocity of 2.94 m/s, the largest among the ISWs in global oceans [9]. Cai et al. summarized the generation and evolution of ISWs in the South China Sea [10].

Sound propagation is the most effective method of information transmission in the ocean, and ISWs have a great impact on underwater sound propagation because they cause time-dependent spatial variations in the water temperature and sound speed [11–14]. The drastic variations in the sound propagation characteristics under ISWs can cause changes in arrival time, propagation path, transmission loss, horizontal refraction and other factors of acoustic waves [15,16]. Thus, ISWs are potentially destructive, decreasing the detection performance of sonar systems.

In addition to sound propagation characteristics, ISWs have been observed to contribute substantially to ambient noise by breaking waves on the sea surface, increasing the internal velocity of currents, and stirring marine sediments [12]. There are three types of noise induced by ISWs, which have been widely studied.

The first type of noise can be easily observed, because the emergence of ISWs forms intense surface rips and produces a noise that is identifiable to the human ear [17]. The collapse of surface waves brings air into the sea and forms a large number of bubbles. In the process of bubble growth and rupture, narrow-band pulses are emitted near the resonant frequency [18]. Rip-band noise can increase the ambient noise by 18 dB at frequencies of 5–15 kHz [17,18]. Near-bottom currents induced by ISWs generate the second type of noise by moving sediments [12,19]. After the sediment particles have been moved away from the seabed, they produce noise by colliding with each other and with the shells of near-bottom hydrophones. Sediment-generated noise is present at frequencies above 10 Hz, but a spectral maximum occurs between 2 kHz and 10 kHz [20–22].

When ISWs flow past a hydrophone and an entire mooring system, pressure fluctuations occur in the turbulence [12]. This noise is referred to as flow noise (the third type), usually below 100 Hz [23]. The intensity of flow noise decreases with increasing frequency. It is usually observable on a mooring-mounted hydrophone array in the cable, but not easily detected on a fixed hydrophone on the seabed [23]. Research on flow noise has a long history. As early as 1960, Willis and Dietz measured the flow noise associated with the tides at frequencies of 40–100 Hz in Narragansett [24,25]. Deane found that the flow noise of mooring instruments in shallow water was mostly below 50 Hz [26]. Strasberg and Webb proposed an infrasonic flow-noise model, but there is as yet no generalized flow-noise model for frequencies exceeding the infrasonic range (frequencies above 20 Hz) [27,28]. Flow noise is closely associated with the sensor size and the entire mooring system.

There have been a number of works on the effects that internal waves have on underwater noise [11–28]. However, noise induced by ISWs in the South China Sea has rarely been studied directly—which is also true of the mechanism of the noise. The strong noise caused by ISWs can drown the underwater acoustic signal and reduce signal-to-noise ratio. Therefore, studies of ISW noise would be of special interest to underwater continuous acoustic communication and monitoring with sonar system.

In this study, we observed the water temperature and flow velocity changes induced by ISWs as part of the 2016 Internal Solitary Wave Cooperative Observation Experiment in the South China Sea. Then, we analyzed the effects of ISWs on noise using acoustic data from a mooring-mounted hydrophone array. We found that the low-frequency flow noise was produced by the cable vibrating under the impact of ISWs, named vortex-induced vibration (VIV). This paper has been organized as follows. The experiment and data are introduced in Section 2. Analysis of noise induced by ISWs are presented in Section 3. A comparison between ISW observations in the 2016 experiment and the 2019 experiment in the South China Sea is provided in Section 4. Finally, Section 5 gives a conclusion.

2. Experiment and Data

In this work, data were collected during the 2016 Internal Solitary Wave Cooperative Observation Experiment. The experiment was conducted in the north of the South China Sea from July 2016 to July 2017. Cooperative observation, consisting of oceanographic and acoustic instrumentation, was adopted in the experiment. A subsurface mooring system was deployed at a station labeled IW5 (117.87° E, 21.11° N) with a water depth of 1000 m, which was shown in Figure 1. Oceanographic data measured by thermistor chains and the acoustic doppler current profilers (ADCPs) were continuously observed. Both upward and downward-looking ADCPs were mounted on the mooring system at depth of 511 m in order to record flow velocity information. The temporal resolution was 3 min, and the vertical resolution was 16 m, covering a depth range of 110~850 m. The mooring system was equipped with thermistor chains of temperature loggers and Conductivity-Temperature-Depth (CTD) recorders between 110–930 m to collect temperature and salinity data. The temporal resolution was also 3 min.

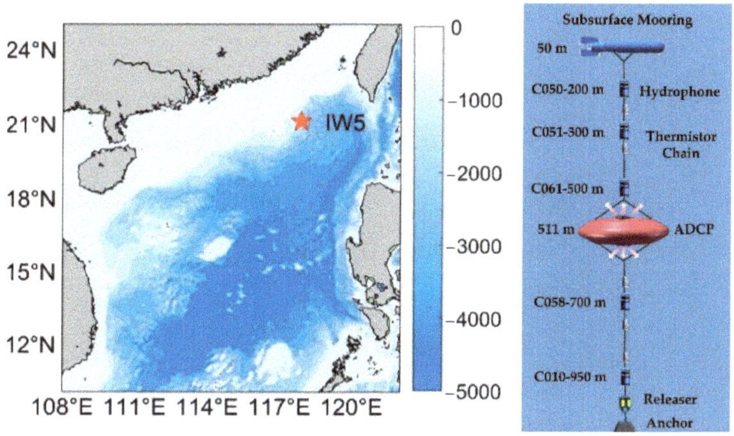

Figure 1. The topography in the South China Sea. The red pentagram indicates the location of the experimental site (IW5 station). The right figure shows the subsurface mooring system.

The acoustic data were measured by a mooring-mounted hydrophone array attached on the cable. The array consisted of five independent acoustic hydrophones, spanning a depth of 200–950 m (Figure 1). The commercial hydrophones used on the array were independent acoustic hydrophones manufactured by the Institute of Acoustics of the Chinese Academy of Sciences. All hydrophones on the array were calibrated before experiment. Their received voltage responses were characterized. The sampling rate of the hydrophone was 4 kHz, and the sound pressure sensitivity of the hydrophone was −175 dB. The sensitivity value was fulfilled between 8 Hz to 2000 Hz without decaying more than −2 dB. Considering the storage and device power, the hydrophone was operated in intermittent mode (15 min every 3 h).

ISWs can cause large depressions on the isothermal surface and notable increases in the flow velocity. As shown in Figure 2, the duration time of temperature and velocity segments was 1 h, spanning a depth of 110−850 m. The influence time of ISWs on seawater temperature was about 15 min. The amplitudes of different ISWs varied greatly, from approximately 90 m to 160 m. The characteristic data for the three different ISWs focused in the study are shown in Table 1.

Table 1. Characteristic data for ISWs on three different days.

Observation Date	Time (UTC+8)	Max Amplitude	Max Flow Velocity
20170314	18:06	159.65 m	−2.09 m/s
20161001	18:03	118.28 m	−1.78 m/s
20160921	05:54	98.30 m	−1.52 m/s

The mooring-mounted hydrophone array and thermistor chains were configured together. It took more than ten minutes for the internal wave to pass the mooring system, so we needed to find the noise data, with internal waves occurring and acoustic records, for analysis. In spite of the experiment period of one year, we only found a total of 6 events that matched the requirement, all of which were analyzed. Three with large internal wave amplitudes were selected for analysis, and the rest of the data also showed similar phenomena (Figure 2). Figure 3 depicts the data received by the C058 hydrophone at a depth of 700 m between 18:00 UTC+8 to 18:15 UTC+8 on 1 October 2016. Figure 3a shows the time-domain signal and the obvious noise burst when the internal wave arrived at 18:03 UTC+8. The time–frequency spectrum (Figure 3b) was obtained by short-time Fourier transform of the data. This analysis used a Hamming window of length 4096, with a 4096-point fast Fourier transform (FFT) and an overlap of 50%. The label of the time–frequency spectrum was power spectral density (PSD). Noise at about 500 Hz and some high frequency lines were generated by the shaking of the hydrophone and the collision of suspended particles in the water against the hydrophone casing when ISWs flowed past the mooring system.

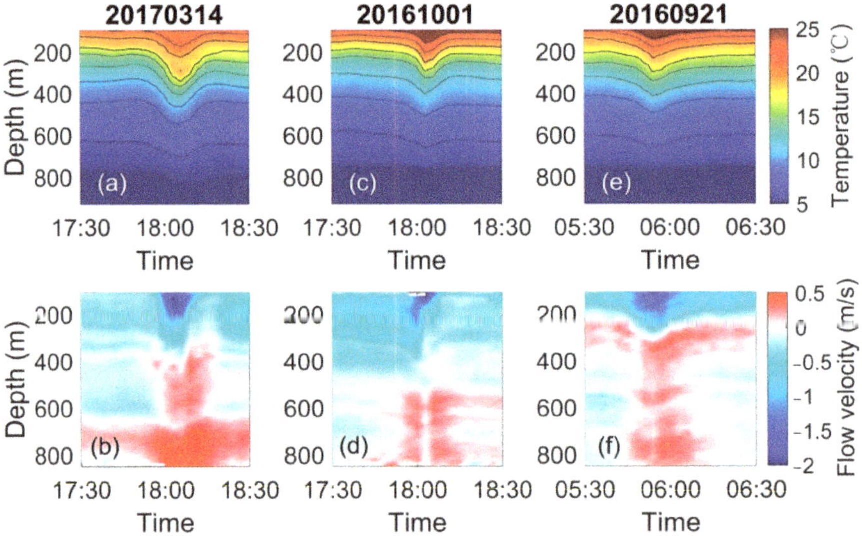

Figure 2. Temperatures and flow velocities for ISWs passing by the IW5 station on three different days. (**a**,**c**,**e**) Temperatures measured by thermistor chains. (**b**,**d**,**f**) Flow velocities measured by the ADCPs. The flow velocity from west to east is positive and the reverse is negative. The flow velocity in this work refers to the zonal velocity. All data sampled at UTC+8.

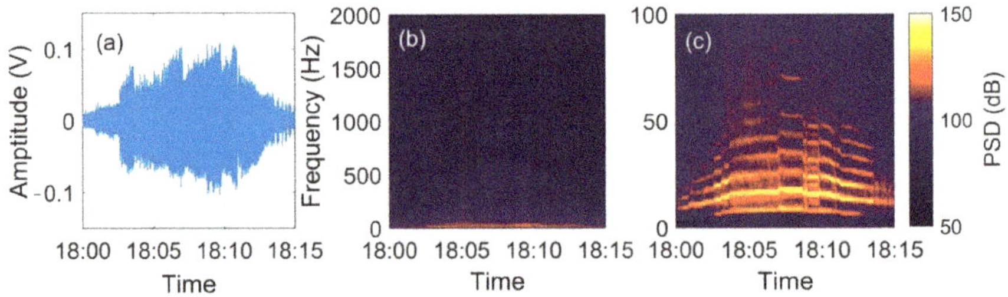

Figure 3. The signal received by the C058 hydrophone at a depth of 700m from 18:00 UTC+8 to 18:15 UTC+8 on 1 October 2016. (**a**) Time–domain signal. (**b**) Time–frequency spectrum. (**c**) Time–frequency spectrum between 0 Hz and 100 Hz.

It was notable that the low-frequency noise presented an identifiable frequency fluctuation at frequencies below 100 Hz, and the acoustic intensity increased with decreasing frequency. A further finding was that low-frequency noise had uniform harmonics, as depicted in Figure 3c (bright stripes). The generation mechanism for these features is discussed in the next section.

3. Analysis and Discussion of Noise Induced by Internal Solitary Waves

3.1. Korteweg-de Vries Equation Theory

The interfacial waves in an arbitrarily stratified fluid can be expressed by the classical Korteweg–de Vries (KdV) equation

$$\frac{\partial \eta}{\partial t} + C_0 \frac{\partial \eta}{\partial x} + \alpha \eta \frac{\partial \eta}{\partial x} + \beta \frac{\partial^3 \eta}{\partial x^3} = 0 \tag{1}$$

where η characterizes the vertical displacement of the isopycnal surface [15,29]. C_0 is a linear speed. The nonlinear parameter α and dispersion parameter β are considered environmental parameters and make contributions to density stratification and sheer currents [10].

In the two-layer fluid model, the upper layer has a thickness h_1 and a density ρ_1, and the lower layer has a thickness h_2 and a density ρ_2, and $\rho_2 > \rho_1$ [15]. This models a typical pycnocline. In this case, the linear speed is

$$C_0 = \sqrt{\frac{g h_1 h_2 \delta\rho}{\rho_{av}(h_1 + h_2)}}, \tag{2}$$

where $\rho_{av} = (\rho_1 + \rho_2)/2$ is the mean density, $\delta\rho = \rho_2 - \rho_1$ is the density difference.

To illustrate the dynamics of a single wave packet of ISW in Figure 2, the solitary solution of KdV equation is

$$\eta = \eta_0 \mathrm{sech}^2 \frac{x - C_p t}{l}, \tag{3}$$

where η_0 is the amplitude of the displacement, x is the horizontal position of an one-dimensional interfacial wave, and the nonlinear velocity C_p and soliton width l [15,29].

$$C_p = C_0 + \frac{\alpha \eta_0}{3}, \; l = \sqrt{\frac{12\beta}{\alpha \eta_0}} \tag{4}$$

The nonlinear and dispersion parameters (α and β) of the model are

$$\alpha = \frac{3C_0(h_1 - h_2)}{2h_1 h_2}, \; \beta = \frac{C_0 h_1 h_2}{6}. \tag{5}$$

Using the explicit form of environmental parameters α and β, Equation (4) becomes

$$C_p = \left[1 + \frac{\eta_0(h_1 - h_2)}{2h_1 h_2}\right], \; l = \frac{2h_1 h_2}{\sqrt{3\eta_0 |h_1 - h_2|}}. \tag{6}$$

This solitary solution is the typical characteristic of ISW in many areas of the ocean [15].

3.2. Spectra Comparison of ISW Noise and Ambient Noise

This section served to illustrate the effects of ISW noise on the ambient noise. Figure 4a shows the power spectra of the noise received by the hydrophone located at 700 m during the passage of internal waves and the ambient noise received three hours later (from 9:00 to 9:02 UTC+8). The data duration was 2 min. The differences in PSD between the two PSD curves are shown in Figure 4b. The spectral level of ISW noise was about 20–40 dB higher than that of ambient noise at frequencies between 10 Hz and 50 Hz. Due to the effects of low frequency harmonics, the maximum difference was 47.7 dB. As the frequency increased, the spectral level gap narrowed. Up to 2 kHz, the difference remained around 10 dB. Flow noise was the dominant source below 100 Hz. Distant ships may be responsible for the peaks in the ambient noise between 200 Hz and 500 Hz. The noise comparison revealed that the ambient noise was completely drowned out by the ISW noise below 2 kHz.

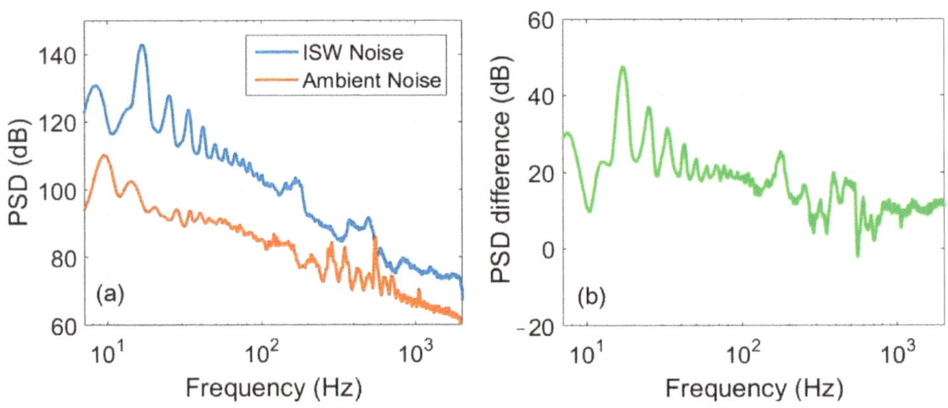

Figure 4. Power spectrum comparison of ISW noise and ambient noise. (**a**) PSD comparison. (**b**) The differences of PSD between ISW noise and ambient noise. ISW noise is from the 2 min data received by hydrophone C058 at depth of 700 m between 6:00 to 6:02 UTC+8 on 21 September 2016, and the ambient noise is from data three hours later (from 9:00 to 9:02 UTC+8).

3.3. Low-Frequency Noise Induced by ISWs

3.3.1. Relationship between Low-Frequency Noise and ISWs

We interpreted the observed low-frequency noise as the flow noise induced by ISWs [23,27]. For this purpose, we analyzed data collected from other ISWs. Figure 5 shows the time–frequency spectra of low frequency noise and their corresponding flow velocities (The data time is consistent with Table 1). The increase in flow velocity was almost synchronous with the burst of low frequency noise. The noise on 21 September 2016 was incomplete, as the maximum velocity occurred at 5:54 UTC+8. As discussed, these results were in accordance with the properties of flow noise induced by ISWs.

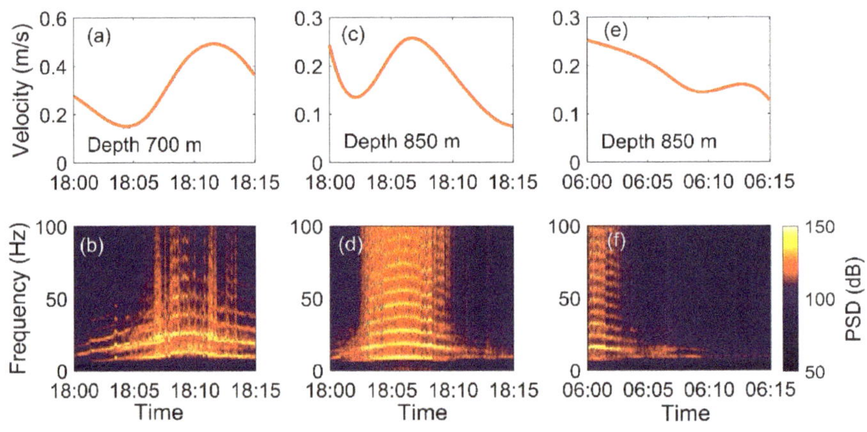

Figure 5. Flow velocities and time–frequency spectra (dB) of the C058 hydrophone (700 m) on (**a**,**b**) 14 March 2017. Flow velocities and time–frequency spectra of the C010 hydrophone (950 m) on (**c**,**d**) 1 October 2016 and (**e**,**f**) 21 September 2016. The depth on the figures represents the depth at which the flow velocity was measured.

It was interesting to note that each segment of noise induced by the ISWs had fundamental and harmonic waves. To elucidate the characteristics of the harmonic waves, we used a 16.384 s data block with 50% overlap (65536-points FFT) from the center of each signal for spectrum analysis. The Welch method and Blackman window were used to calculate power spectra. Figure 6 shows the power spectra corresponding to the three signals in Figure 5. Table 2 shows their fundamental and harmonic frequencies. Each segment of signal had a different fundamental frequency. This phenomenon may be associated with the velocity of the ISW. It was worth discussing these facts, in terms of which part of the mooring system the internal wave interacted with, to produce low-frequency flow noise and harmonics.

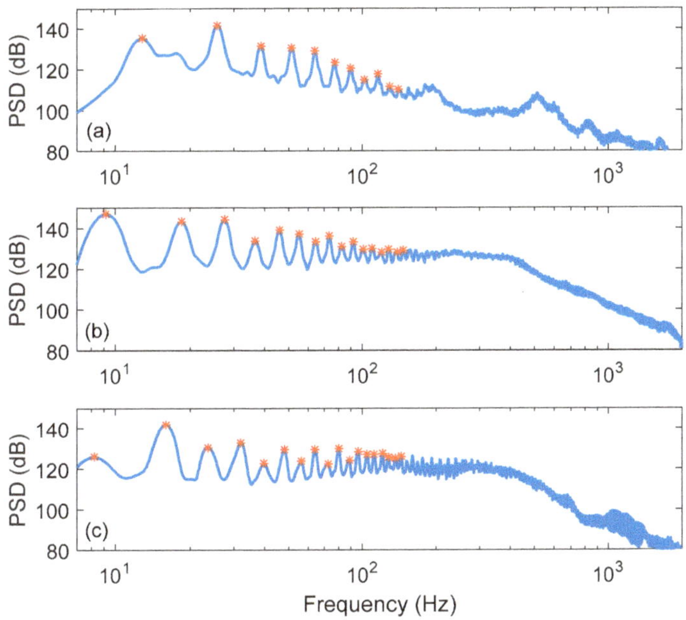

Figure 6. Power spectra of the noise recorded by the same hydrophones as in Figure 5. (**a**) The data after 18:09 UTC+8. (**b**) The data after 18:06 UTC+8. (**c**) The data after 6:00 UTC+8. The data length is 16.384 s.

Table 2. The fundamental and harmonic frequencies of the three signals (Hz).

Fundamental	Double	Triple	Quadruple	Five-Times	Six-Times	Seven-Times	Eight-Times
12.72	25.39	39.06	51.76	64.65	77.15	89.84	105.32
9.16	18.43	27.59	36.74	45.89	55.05	64.58	73.49
8.18	16.11	24.17	32.35	40.41	48.58	56.52	64.58

3.3.2. Vortex-Induced Vibration

The cable of the mooring system can be abstracted as a flexible riser in the water with the upper end subjected to vertical tension and the lower end fixed. The current, when flowing past the cylinder, generates a vortex alternately, therefore, periodic pressure changes occur around the cylinder [30]. When the cylinder is a flexible structure, the periodic vibration of the cylinder structure is caused by the alternating pressure, namely vortex-induced vibration [30]. This can be used to account for the vibration caused by internal waves flowing past the cable system.

As far as we know, little previous research has investigated the VIV induced by ISWs. However, VIV produced by other kinds of currents around a cylinder has been widely documented. In 1911, Karman conducted a pioneering study on vortex shedding around the flow based on the non-viscous method. When a viscous fluid passes through a cylindrical structure, its wake will undergo flow separation and form two rows of vortices alternately, which is called a Karman vortex street [31]. When the frequency of the vortex is close to the natural frequency of the structure, structural resonance or VIV occurs. Kassen [32], Sarpkaya [33] and Wu [34] provided detailed explanations of VIV, including generation mechanism, models, wake form, self-locking phenomenon and so on.

Although there have been many studies of VIV on underwater flexible risers in the laboratory, there have been few underwater experimental studies owing to the high cost of full-scale field experiments. Large-scale model tests of tensioning risers (90 m in length and 3 cm in diameter) were carried out in a lake on Norway's west coast in 1997 [35]. Lie used modal analysis to determine that the VIV was irregular in the non-lock-incase, and that the vibration frequency was composed of the natural frequency of the structure plus the frequency of vortex shedding along the length of the structure [35]. Recent studies have shown that the VIV of the flexible riser had multi-order harmonics. Wu determined that higher harmonics could be present with flexible beam VIV [36]. Trim analyzed the second harmonic frequency component of the VIV from a flexible riser (38 m in length and 2.7 cm in diameter) in the Marintek Ocean Basin in Trondheim [37]. When Wang analyzed the multi-order modes and harmonics of the cable-coupled vibration response, he found that, the closer to the midpoint of the cable, the higher the excitation degree of the first two modes, and the farther away from the midpoint of the cable, the more average the excitation degree of the first six modes. [38]. The uniform harmonic component was evidence of internal wave noise. This was also consistent with the results of our analysis in Figure 6. It was considered that the oscillations of the marine cables system would cause noise in the sonar system, reducing the performance of other environmental sensors [30].

When uniform turbulence flows past a rigid, fixed riser of diameter d, the time-varying motion of the riser can be made up of a series of mode-shapes if the dynamical process of the riser vibration is approximately linear [35].

$$x(t,z) = \sum_{n=1}^{\infty} w_n(t)\varphi_n(z),\ z \in [0, L], \tag{7}$$

where L is the length of the riser, z is the vertical coordinate, and t is the time. The horizontal displacement of the riser $x(t,z)$ is determinded by the mode-shape $\varphi_n(z)$ and the modal weight $w_n(t)$, $n = 1, 2, 3 \ldots$ [35].

The vortex shedding frequency is

$$f_{VIV} = St \cdot u/d, \tag{8}$$

where the u is the free-current velocity. When the cable diameter is no more than 0.1 m, the maximum Reynolds number for the mooring system is usually less than 10^5, and during a wide range of the Reynolds number (Re), $10^2 < \text{Re} < 10^5$, the Strouhal number (St) varies little and has a value around 0.2 [30,35].

We knew the diameter of the cable (1 cm) and the velocity at the depth of the hydrophones at different times. The vortex shedding frequency was consistent with the structural natural frequency [30]. The shedding frequency was calculated using Equation (8). As shown in Table 3, shedding frequency was positively correlated with the ISW velocity. The velocity increased as the internal wave approached the hydrophone and decreased as the internal wave moved away from the hydrophone. Therefore, the frequency of the noise appearred to increase first and then decrease, appearing as a frequency fluctuation in the time–frequency spectra.

Table 3. ISW velocities and calculated frequencies of the hydrophones at different times.

Hydrophone	C058-20170314 (700 m)		C010-20161001 (950 m)		C010-20160921 (950 m)	
Time	18:03	18:09	18:03	18:06	06:00	06:03
Velocity (cm/s)	41.82	52.90	21.74	25.26	23.62	17.11
Frequency (Hz)	8.36	10.58	4.35	5.05	4.72	3.42

The fundamental frequencies (Table 2) of the noise with maximum spectra at 18:09, 18:06 and 06:00 UTC+8 were 12.72 Hz, 9.16 Hz and 8.18 Hz, respectively. The actual frequency was higher than the calculated frequency. The vibration of the cable depended not only on the velocity of the internal waves, but also on the mass of the cable, the added mass, the elasticity and the damping coefficient [35]. Locking and synchronization may occur if the natural resonant frequency of the cable approximates the Strouhal frequency [30]. In this case, the vortex fell off at the actual frequency, not the frequency calculated by Equation (8), because the formation and shedding process of the vortex changed the added mass of the cable [30,35]. The change in added mass may be positive or negative, causing an increase or decrease in shedding frequency [35]. In addition, the velocity of ISW varied greatly with time and space, further complicating the vortex shedding frequency [38].

Together, the characteristics and frequency components of the VIV were consistent with our analysis of flow noise. The results of the experiment found clear support for low frequency noise induced by VIV when ISWs flowed past the marine cable system.

3.4. Spectrum Comparison of ISW Noise at Different Depths

To illustrate the depth dependence of ISW noise, Figure 7 shows the power spectra comparison as a function of the frequency for five hydrophones at depths of 200 m, 300 m, 500 m, 700 m, and 950 m. As shown in Figure 7, the distant noise component below 100 Hz was derived from the flow noise induced by ISWs. The fundamental frequencies of the noise differed substantially because of the velocity difference of internal waves at different depths.

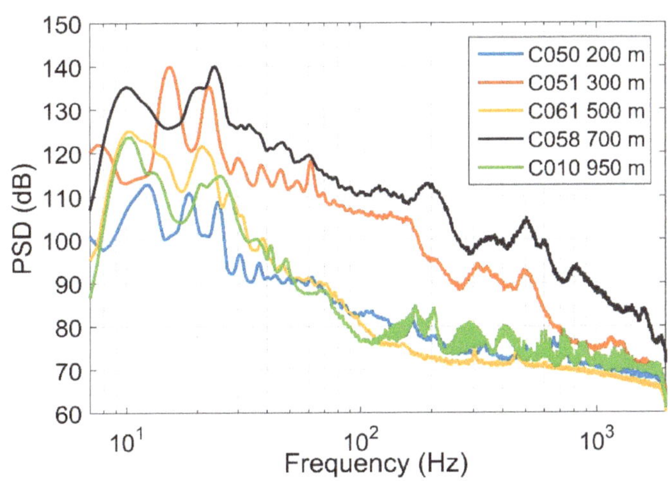

Figure 7. Power spectra comparison of ISW noise from five hydrophones at different depths from 18:06 to 18:08 UTC+8 on 14 March 2017.

The noise spectrum levels of the hydrophones at 200 m, 500 m, and 700 m were less than those at 300 m and 700 m. The key was the structure of the entire mooring system. The vibration amplitude of the cable decreased with the increase of tension [38,39]. The hydrophone, C050, at 200 m, was close to the main floater at the top of the cable system.

Hydrophone C061, at 500 m, approached the central floater where the ADCPs were located. Hydrophone C010, at 950 m, was located near the fixed object at the bottom of the cable. Hydrophone C058 had a larger spectrum level than hydrophone C051, because the floater at 500 m divided the cable into two sections, shorter at the top and longer at the bottom. The longer the vibration length of the cable, the greater the vibration amplitude [39].

4. Comparison of ISW Noise in 2019 ISW Observation Experiment

This section compares noise measurements from the 2019 Internal Solitary Wave Observation Experiment. The experiment was located on the continental shelf of the South China Sea with an average water depth of 360 m.

Figure 8 shows the temperature data and the noise of the same internal wave received by two hydrophones. The maximum amplitude of this ISW was 91.74 m which arrived at 19:42 UTC+8. Figure 8b shows the time–frequency spectrum below 100 Hz recorded by a hydrophone located at depth of 200 m, which was the middle hydrophone of the mooring-mounted linear array. The length of the cable in this experiment was about 300 m. The fundamental frequency of low frequency noise was 10.20 Hz, with uniform harmonic components.

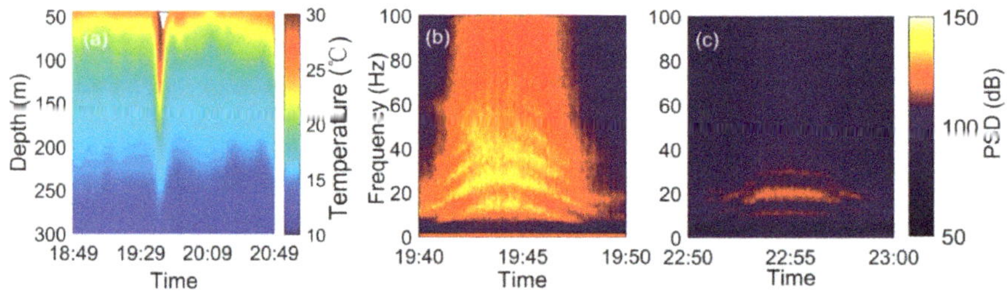

Figure 8. Temperature data and noise of the same ISW received by two hydrophones on 9 July 2019. (**a**) Temperature measured by the thermistor chains. (**b**) Time–frequency spectrum from a hydrophone at a depth of 250 m. The cable of the hydrophone array is 300 m long. (**c**) Time–frequency spectrum from the near–bottom hydrophone at a depth of 340 m. The cable of the single hydrophone is 30 m long.

The results confirmed that the flow noise induced by ISWs was indeed caused by the VIV of the cable system. The noise in Figure 8c was from the near-bottom hydrophone at a depth of 340 m, 17.8 km west of the mooring-mounted linear array. The length of the cable that held the hydrophone was 30 m. When the same ISW flowed past the near-bottom hydrophone, the noise intensity was substantially reduced. The fundamental frequency of low frequency noise was 6.15 Hz.

A comparison between ISW observations in the 2016 experiment and the 2019 experiment is shown in Table 4. The amplitude of the internal wave in the 2019 experiment was slightly smaller than that observed in the 2016 experiment. Due to the shallow sea, the cable length of the mooring system was much smaller than that of the deep-sea experiment. This result showed that the noise intensity induced by ISW was related to the cable length, because the longer the cable, the stronger the VIV [30,35]. This was also related to the position of the hydrophone attached on the mooring system. Overall, all noise datasets had uniform harmonics, which was in accordance with the characteristics of flow velocity.

Table 4. The comparison between ISW observations in the 2016 experiment and 2019 experiment.

	2019 Experiment		2016 Experiment		
Water depth (m)	360 m		1000 m		
Cable length (m)	300 m	30 m	490 m	490 m	490 m
Amplitude of ISW (m)	91.74 m	91.74 m	159.65 m	118.28 m	98.30 m
Hydrophone location	middle	bottom	middle	bottom	bottom
Fundamental frequency (Hz)	10.20 Hz	6.15 Hz	12.72 Hz	9.16 Hz	8.18 Hz

5. Conclusions

In this study, oceanographic and acoustic data from the 2016 Internal Solitary Wave Cooperative Observation Experiment in the South China Sea were studied. We analyzed the noise during three large ISWs and found that ISWs produced strong noise at low frequencies. The analysis results could be used as a supplement to internal wave observations and provide a passive acoustic monitoring and warning method for high marine currents.

Within 15 min of the internal wave flowing past the subsurface mooring, the isothermal surface was depressed substantially, with a maximum amplitude reaching 160 m and the maximum velocity exceeding 2 m/s. The noise received by the mooring-mounted hydrophone array also increased.

Compared with the ambient noise, the results indicated that the ISW noise was higher than the ambient noise at frequencies below 2 kHz, and even higher than 20–40 dB within 100 Hz.

Through analysis of time–frequency spectra and power spectra, we interpreted the low frequency noise as vortex-induced vibration caused by ISWs flowing past the marine cable system owing to the significant harmonic component of the noise. The noise duration was basically consistent with the time of ISWs interacting with the mooring system. It was notable that the frequency fluctuation of noise was caused by the variation of internal wave velocities with time and space. Through comparison of the ISW noises received by hydrophones at different depths, we found that the power spectra of ISW noises were associated with the positions of the hydrophones fixed on the cable. The closer to the fixed end, the smaller the noise spectrum level, and vice versa.

For further evidence, the observation experiment of internal waves in the South China Sea in 2019 confirmed that the frequency fluctuation and harmonic components of low frequency noise induced by internal solitary waves still occurred under different experimental conditions.

The results of this study could establish a high marine current warning and real-time marine current monitoring system, because the observed acoustic characteristics, including harmonics, frequency fluctuation, and energy enhancement, were intrinsically related to the velocity and amplitude of internal waves. Oceanographic phenomena can be studied based on acoustic observations and noise analysis. In addition, when studying oceanographic phenomena, acoustic signal analysis technology can receive data through a single hydrophone on the mooring system, without the full ocean depth thermistor chains and ADCP. It would be both cheaper and more practical than other ocean instruments.

To our knowledge, few studies have examined the model of the marine cable system vibration under ISWs. Therefore, the low frequency noise induced by ISWs calls for further experimentation and theoretical exploration.

Author Contributions: Conceptualization, J.L. and Y.S.; methodology, J.L. and Y.S.; formal analysis, J.L., Y.S., Y.Y. and X.H.; resources, Y.S. and X.H.; data curation, J.L.; writing—original draft preparation, J.L.; writing—review and editing, J.L., Y.S., Y.Y. and X.H.; supervision, Y.Y.; funding acquisition, Y.S. and Y.Y. All authors have read and agreed to the published version of the manuscript.

Funding: This research was funded by the National Natural Science Foundation of China, grant number 41906160, 11974286, 12174312.

Institutional Review Board Statement: Not applicable.

Informed Consent Statement: Not applicable.

Data Availability Statement: The data from 2016 Internal Solitary Wave Cooperative Observation Experiment presented in this study are available on request from the corresponding author.

Acknowledgments: We are grateful to Bingyong Guo for providing helpful suggestions about VIV in the manuscript.

Conflicts of Interest: The authors declare no conflict of interest.

References

1. Helfrich, K.R.; Melville, W.K. Long Nonlinear Internal Waves. *Ann. Rev. Fluid Mech.* **2006**, *38*, 395–425. [CrossRef]
2. Osborne, A.; Burch, T. Internal solitons in the Andaman Sea. *Science* **1980**, *208*, 451–460. [CrossRef] [PubMed]
3. Jackson, C. Internal wave detection using the Moderate Resolution Imaging Spectroradiometer (MODIS). *J. Geophys. Res.* **2007**, *112*, C10012. [CrossRef]
4. Zhang, S.; Alford, M.H.; Mickett, J.B. Characteristics, generation and mass transport of nonlinear internal waves on the Washington continental shelf. *J. Geophys. Res. Oceans* **2015**, *120*, 741–758. [CrossRef]
5. Scotti, A.; Beardsley, R.C.; Butman, B. Generation and propagation of nonlinear internal waves in Massachusetts Bay. *J. Geophys. Res.* **2007**, *112*, C10001. [CrossRef]
6. Lee, S.W.; Nam, S. Estimation of Propagation Speed and Direction of Nonlinear Internal Waves from Underway and Moored Measurements. *J. Mar. Sci. Eng.* **2021**, *9*, 1089. [CrossRef]
7. Huang, X.; Chen, Z.; Zhao, W.; Zhang, Z.; Zhou, C.; Yang, Q.; Tian, J. An extreme internal solitary wave event observed in the northern South China Sea. *Sci. Rep.* **2016**, *6*, 30041. [CrossRef]
8. Ramp, S.R.; Tang, T.Y.; Duda, T.F.; Lynch, J.F.; Liu, A.K.; Chiu, C.S.; Bahr, F.L.; Kim, H.R.; Yang, Y.J. Internal Solitons in the Northeastern South China Sea Part I: Sources and Deep Water Propagation. *J. Ocean. Eng.* **2004**, *29*, 1157–1181. [CrossRef]
9. Xu, A.; Chen, X. A Strong Internal Solitary Wave with Extreme Velocity Captured Northeast of Dong-Sha Atoll in the Northern South China Sea. *J. Mar. Sci. Eng.* **2021**, *9*, 1277. [CrossRef]
10. Cai, S.; Xie, J.; He, J. An overview of internal solitary waves in the South China Sea. *Surv. Geophys.* **2012**, *33*, 927–943. [CrossRef]
11. Colosi, J.A.; Scheer, E.K.; Flatte, S.M.; Cornuelle, B.D.; Dzieciuch, M.A.; Munk, W.H.; Worcester, P.F.; Howe, B.M.; Mercer, J.A.; Spindel, R.C.; et al. Comparisons of measured and predicted acoustic fluctuations for a 3250-km propagation experiment in the eastern North Pacific Ocean. *J. Acoust. Soc. Am.* **1999**, *105*, 3202–3218. [CrossRef]
12. Katsnelson, B.G.; Godin, O.A.; Zhang, Q. Observations of acoustic noise bursts accompanying nonlinear internal gravity waves on the continental shelf off new jersey. *J. Acoust. Soc. Am.* **2021**, *149*, 1609–1622. [CrossRef]
13. Simmen, J.; Flatte, S.M.; Wang, G.Y. Wavefront folding, chaos, and diffraction for sound propagation through ocean internal waves. *J. Acoust. Soc. Am.* **1997**, *102*, 239–255. [CrossRef]
14. Katsnelson, B.G.; Grigorev, V.; Badiey, M.; Lynch, J.F. Temporal sound field fluctuations in the presence of internal solitary waves in shallow water. *J. Acoust. Soc. Am.* **2009**, *126*, EL41–EL48. [CrossRef]
15. Apel, J.R.; Ostrovsky, L.A.; Stepanyants, Y.A.; Lynch, J.F. Internal solitons in the ocean and their effect on underwater sound. *J. Acoust. Soc. Am.* **2007**, *121*, 695–722. [CrossRef]
16. Badiey, M.; Mu, Y.; Lynch, J.; Apel, J.; Wolf, S. Temporal and azimuthal dependence of sound propagation in shallow water with internal waves. *IEEE J. Ocean. Eng.* **2002**, *27*, 117–129. [CrossRef]
17. Tang, D.; Moum, J.N.; Lynch, J.F.; Abbot, P.; Chapman, R.; Dahl, P.H.; Duda, T.F.; Gawarkiewicz, G.; Glenn, S.; Goff, J.A.; et al. Shallow Water '06: A joint acoustic propagation/nonlinear internal wave physics experiment. *Oceanography* **2007**, *20*, 156–167. [CrossRef]
18. Wei, R.C.; Chien, K.F.; Chiu, L.; Chen, C.F. Analysis of low frequency ocean ambient noise induced by internal wave in sand dune region of northern South China Sea. *J. Acoust. Soc. Am.* **2016**, *140*, 3013. [CrossRef]
19. Serebryany, A.N.; Newhall, A.; Lynch, J.F. Observations of noise generated by nonlinear internal waves on the continental shelf during the SW06 experiment. *J. Acoust. Soc. Am.* **2008**, *123*, 3589. [CrossRef]
20. Yang, Y.J.; Chiu, C.S.; Wu, J.C.; Liang, W.D.; Ramp, S.R.; Reeder, D.B.; Chen, C.F. Observations of ambient noises induced by the internal solitary waves on the continental slope of the northern South China Sea: Ambient noises by ISW in SCS. In Proceedings of the OCEANS 2015-Genova, Genova, Italy, 21 September 2015; pp. 1–4.
21. Bassett, C.; Thomson, J.; Polagye, B.L. Sediment-generated noise and bed stress in a tidal channel. *J. Geophys. Res. Oceans* **2013**, 2249–2265. [CrossRef]
22. Bourgault, D.; Morsilli, M.; Richards, C.; Neumeier, U.; Kelley, D.E. Sediment resuspension and nepheloid layers induced by long internal solitary waves shoaling orthogonally on uniform slopes. *Cont. Shelf Res.* **2014**, *72*, 21–33. [CrossRef]
23. Bassett, C.; Thomson, J.; Dahl, P.H.; Polagye, B. Flow-noise and turbulence in two tidal channels. *J. Acoust. Soc. Am.* **2014**, *135*, 1764–1774. [CrossRef]
24. Dietz, F.T.; Kahn, J.S.; Birch, W.R. Nonrandom associations between shallow water ambient noise and tidal phase. *J. Acoust. Soc. Am.* **1960**, *32*, 915. [CrossRef]
25. Willis, J.; Dietz, F.T. Some characteristics of 25-cps shallow-water ambient noise. *J. Acoust. Soc. Am.* **1965**, *37*, 125–130. [CrossRef]

26. Deane, G.B. Long time-base observations of surf noise. *J. Acoust. Soc. Am.* **2000**, *107*, 758–770. [CrossRef]
27. Strasberg, M. Non-acoustic noise interference in measurements of infrasonic ambient noise. *J. Acoust. Soc. Am.* **1979**, *66*, 1487–1493. [CrossRef]
28. Webb, S.C. Long-period acoustics and seismic measurements and ocean floor currents. *J. Ocean. Eng.* **1988**, *13*, 263–270. [CrossRef]
29. Gerkema, T. *An Introduction to Internal Waves*; Lecture Notes; Royal NIOZ: Texel, The Netherlands, 2008; pp. 147–164.
30. Every, M.J.; King, R.; Weaver, D.S. Vortex-excited vibrations of cylinders and cables and their suppression. *Ocean Eng.* **1982**, *9*, 135–157. [CrossRef]
31. Pankanin, G.L.; Kulińczak, A.; Berliński, J. Investigations of Karman vortex street using flow visualization and image processing. *Sensors Actuat. Phys.* **2007**, *138*, 366–375. [CrossRef]
32. Kaasen, K.E.; Lie, H.; Solaas, F.; Vandiver, J.K. Norwegian Deepwater Program: Analysis of Vortex-Induced Vibrations of Marine Risers Based on Full-Scale Measurements. In *Offshore Technology Conference*; OTC Program Committee: Houston, TX, USA, 2000; p. OTC-11997-MS.
33. Sarpkaya, T. Hydrodynamic damping, flow-induced oscillations, and biharmonic response. *J. Offshore Mech. Arct. Eng.* **1995**, *117*, 232–238. [CrossRef]
34. Wu, J.; Yin, D.; Halvor, L. On the occurrence of higher harmonics in the VIV response. In *OMAE 2015 34th International Conference*; AMSE: St. John's, NL, Canada, 2015; p. OMAE2015-42061.
35. Lie, H.; Kaasen, K.E. Modal analysis of measurements from a large-scale VIV model test of a riser in linearly sheared flow. *J. Fluid. Struct.* **2006**, *22*, 557–575. [CrossRef]
36. Trim, A.D.; Braaten, H.; Lie, H.; Tognarelli, M.A. Experimental investigation of vortex-induced vibration of long marine risers. *J. Fluid. Struct.* **2005**, *21*, 335–361. [CrossRef]
37. Wang, J.X.; Gui, H.B.; Chen, X. Analysis of Attitude and Dynamics Characteristic of a Kind of Submerged Buoy. *Appl. Mech. Mater.* **2012**, *226*, 516–520. [CrossRef]
38. Hover, F.S.; Miller, S.N.; Triantafyllou, M.S. Vortex-induced vibration of marine cables: Experiments using force feedback. *J. Fluid. Struct.* **1997**, *11*, 307–326. [CrossRef]
39. Owen, M.G.; Richard, A.S.; Steven, E.R. The Resonant, Vortex-Excited Vibrations of structures and Cable Systems. In *Offshore Technology Conference*; OTC Program Committee: Houston, TX, USA, 1975; p. OTC-2319-MS.

Journal of
Marine Science and Engineering

MDPI

Article

Estimation of Propagation Speed and Direction of Nonlinear Internal Waves from Underway and Moored Measurements

Seung-Woo Lee [1] and Sunghyun Nam [1,2,*]

1 School of Earth and Environmental Sciences, College of Natural Sciences, Seoul National University, Seoul 08826, Korea; lsw.ocean@gmail.com
2 Research Institute of Oceanography, College of Natural Sciences, Seoul National University, Seoul 08826, Korea
* Correspondence: namsh@snu.ac.kr

Abstract: Propagation speed and direction of nonlinear internal waves (NLIWs) are important parameters for understanding the generation and propagation of waves, and ultimately clarifying regional ocean circulation. However, these parameters cannot be directly measured from in-situ instruments, but can only be estimated from post-processing in situ data. Herein, we present two methods and an optimal approach to estimate the propagation speed and direction of waves using underway and moored observations. The *Doppler shift* method estimates these parameters from apparent observations concerning a moving ship using the Doppler shift induced by the changing relative distance of the NLIWs from the moving ship. The *time lag* method estimates the parameters using the distance between two locations of the NLIW observed at different times and the time lag. To optimize the speed and direction of NLIWs, the difference in the propagation direction independently estimated by the two methods needs to be minimized concerning the optimal propagation speed to yield the optimal propagation direction. The methods were applied to two cases observed in the northern East China Sea in May 2015 and August 2018. This study has practical significance for better estimating the propagation speed and direction of NILWs particularly over a broad continental shelf.

Keywords: nonlinear internal wave; propagating speed; propagating direction; underway observation; moored observation; East China Sea

Citation: Lee, S.-W.; Nam, S. Estimation of Propagation Speed and Direction of Nonlinear Internal Waves from Underway and Moored Measurements. *J. Mar. Sci. Eng.* **2021**, *9*, 1089. https://doi.org/10.3390/jmse9101089

Academic Editor: Shuqun Cai

Received: 1 September 2021
Accepted: 2 October 2021
Published: 6 October 2021

1. Introduction

Nonlinear internal waves (NLIWs) are ubiquitous in stratified seas and are accompanied by isopycnal fluctuations with a sharp vertical density gradient. They play an important role in underwater acoustics, regional circulation, local biogeochemistry, and energetics, mostly via vertical mixing in the stages of generation, propagation, evolution, and dissipation. NLIWs affect the transportation of momentum, heat, and energy via turbulent dissipation and mixing [1–5]. Marine ecosystems are significantly influenced by vertical nutrient supply, chlorophyll bloom, and biological redistribution, which can be modulated by NLIWs [6–9]. The NLIWs drive sediment resuspension and transportation; thus, they affect marine geophysics and underwater acoustics [10–15]. Vertical isopycnal displacements, which allow the wave amplitude to be defined, and propagation speed and direction, are fundamental parameters of NLIWs that are useful, but cannot be directly measured from in situ sampling, for a clear understanding of their generation, propagation, evolution, and dissipation. Estimating the propagation speed and direction can be important for assessing regional ocean circulation, biogeochemical cycles, energetics, underwater acoustics, and the dynamics of NLIWs.

Methods to estimate the propagation speed and direction have been suggested but are mostly limited by sampling strategies that have not yet been validated. The most common method using multiple moorings aligned in the propagation direction of NLIWs aims to divide the distance between the mooring locations by the arrival time differences [16–22]. However, it is not practical to deploy many moorings along the ray of NLIWs, particularly where the continental shelf is wide and multiple NLIWs are generated from multiple sources with different unknown propagation directions. Another method is to use the principal direction of the wave-induced horizontal velocity and the temporal difference of enhanced echo intensity from acoustic Doppler current profiler (ADCP) measurements [23–27]. This method is useful, but not very practical, as extracting the propagation speed and direction is not straightforward. Using remote sensors, such as synthetic aperture radar (SAR) and spectroradiometer, the propagation speed and direction can be estimated from the horizontal curvature of satellite images [28–32]; however, the limited spatiotemporal satellite sampling from polar orbits does not allow NLIWs to be easily detected. Therefore, it is necessary to develop a method to estimate the propagation speed and direction of NLIWs from widely used ship-based in situ measurements.

In the northern East China Sea (ECS), NLIWs are mainly formed by strong tidal forces that interact with bathymetric features. NLIWs in this region have been observed in association with strong semidiurnal internal tides over slope areas in the southern and southeastern parts of the ECS [33] and local lee-wave generation by small islands and seamounts near Jeju Island and the Ieodo Ocean Research Station (IORS) in the northern ECS [34–36] (Figure 1). Unlike the typical setting where dominant first-mode NLIWs in a two-layered condition propagate from the shelf break towards the coast, high NLIW modes propagating in multiple directions from multiple sources have been identified in the northern ECS [34,37].

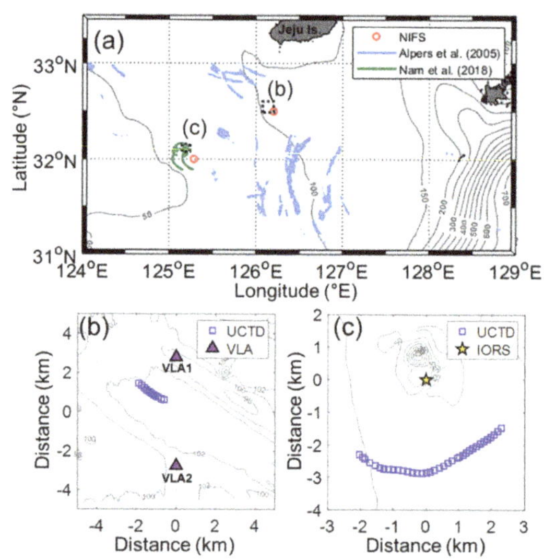

Figure 1. (**a**) Map showing the geographic region of this study for two areas of experiments (black dashed boxes), bathymetry (grey lines), and distribution of surface manifestation of NLIWs described by Alpers et al. [35] (blue lines) and Nam et al. [36] (green lines). Two stations conducting historical hydrographic data sampling for the National Institute of Fisheries Science (NIFS) used in this study are marked by red open circles. Zoomed-in maps of the two areas of (**b**) Shallow-water Acoustic Variability EXperiment 2015 (SAVEX15) and (**c**) Ieodo Ocean Research Station 2018 (IORS18). Locations of underway conductivity–temperature–depth (UCTD) data collection, moored observations (VLA1 and VLA2), and Ieodo Ocean Research Station (IORS) are marked by blue open squares in (**b**,**c**), purple triangles in (**b**), and yellow stars in (**c**).

Herein, we present a new method for estimating the propagation speed and direction of NLIWs using both moored and underway measurements, and the results of applying the method to two cases of NLIWs observed in the northern ECS in May 2015 and August 2018 (Figure 1).

2. Data and Methods

2.1. Data and Processing

Shallow-water Acoustic Variability EXperiment 2015 (SAVEX15) was conducted on 14–28 May 2015, focusing on a relatively small area (water depth: ~100 m, area shown in Figure 1a,b) in the northern ECS [38–42]. During the experiment, two moorings (vertical line arrays [VLAs]) and underway conductivity–temperature–depth (UCTD) instruments were used to collect moored temperature (with no conductivity) time series at multiple depths and vertical profiles [43]. The two moorings (VLA1 and VLA2) deployed at water depths of 101 and 102 m, were horizontally separated by ~5.5 km; 25 temperature loggers and 5 Star-Oddi temperature–depth–tilt sensors were attached at nominal depths of 2–80 m with an interval of 1–5 m. The sampling time interval of the moored temperature sensors was 30 s.

Ieodo Ocean Research Station 2018 (IORS18) was conducted in the northern ECS in the vicinity of IORS (32°7.4′ N, 125°10.9′ E, constructed at a water depth of 41 m) on 28 August–1 September 2018 (Figure 1a,c). During the IORS18 experiment, UCTD was used to collect ship-based vertical profiles of temperature and salinity and the IORS-based time series of temperature and pressure data observed at nominal depths of 2, 5, 11, 16, 22, 32, and 37 m, with a typical sampling time interval of 60 s [44].

The depths of the moored temperature sensors attached to the VLAs were corrected using the tilt and pressure data recorded by five Star-Oddi temperature–depth–tilt sensors. After the removal of outliers, the moored temperature data were vertically interpolated using the Akima spline method [45] at 1 m intervals. The UCTD data were processed following the method described by Ullman and Herbert [46], except for the alignment process to correct the mismatch due to different time delays of the conductivity and temperature sensors. The raw temperature and conductivity measured by the UCTD were filtered with a cut-off period of four scans (0.25 s). The raw pressure measured using the UCTD was filtered with a cut-off period of 32 scans (2 s). The time delay between the conductivity and temperature sensors was corrected using lagged correlation. Then, spikes in the salinity data were removed by aligning the data of the temperature and conductivity sensors. After the alignment processing, the data were corrected for the effect of viscous heating and finally vertically averaged over 1 *dbar* bin.

To discuss the theoretical propagation speed of NLIWs in the northern ECS in the context of long-term and interannual variations, vertical profiles of temperature and salinity routinely observed every other month from 1990 to 2019, at two hydrographic stations of the National Institute of Fisheries Science (NIFS), Republic of Korea, were used in this study (red open circles in Figure 1a). To ensure the quality of the temperature and salinity data, vertical profiles containing unreasonable values (both global and local) were removed. Quality control procedures, such as the spike and gradient tests, were applied to extract reliable salinity and temperature profiles. Because the profiles are only available at standard depths (i.e., surface, 10, 20, 30, 50, 75, 100, 125, 150, 200, and 250 m), linear interpolation was conducted to determine data at 1 m vertical intervals [47].

Moderate resolution imaging spectroradiometer (MODIS) sensors onboard the National Aeronautics and Space Administration (NASA) satellites Terra and Aqua, provided true-color images from calibrated, corrected, and geo-located radiance (Level-1 B) data, with a spatial resolution of 250 m. As NLIWs induce the divergence and convergence of sea surface currents as they propagate, thereby modifying the sea surface roughness, they are visible in MODIS true-color images if they are in a sun-glint area [48]. In this study, two images obtained by MODIS Terra on 2 August 2015, and MODIS Aqua on 30 July 2018, were used to estimate the propagation direction of NLIWs from sea surface manifestations.

2.2. Methods

2.2.1. Two-Layered KdV (Korteweg-de Vries) Theories

In classical KdV theory [49], a leading-order weak non-linearity and dispersion are competing but comparable to each other. For the two-layered KdV theory, the thicknesses (h_1, h_2) and densities (ρ_1, ρ_2) of the upper and lower layers can be used to estimate the parameters of mode-1 NLIWs, including linear phase speed $c_{KdV.l}$, theoretical propagation speed $c_{KdV.iw}$, characteristic width $2\Delta_{KdV.iw}$, nonlinear parameter α, and dispersion parameter β, yielding the wave equation as follows [14,50]:

$$\frac{\partial \eta}{\partial t} + c_{KdV.l}\frac{\partial \eta}{\partial x} + \alpha\eta\frac{\partial \eta}{\partial x} + \beta\frac{\partial^3 \eta}{\partial x^3} = 0, \tag{1}$$

where η, t, and x are the vertical displacement of the isopycnals (or isotherms), time, and horizontal coordinates, respectively. The $c_{KdV.l}$, α, and β can be estimated using the density stratification parameters (ρ_1, ρ_2, h_1, and h_2) in a two-layered system as follows:

$$c_{KdV.l} = \sqrt{g\frac{\rho_2 - \rho_1}{(\rho_1 +]\rho_2)/2}\frac{h_1 h_2}{(h_1 + h_2)}}, \tag{2}$$

$$\alpha = \frac{3}{2}\frac{c_{KdV.l}}{h_1 h_2}\frac{\rho_2 {h_1}^2 - \rho_1 {h_2}^2}{\rho_2 h_1 + \rho_1 h_2}, \tag{3}$$

$$\beta = \frac{c_{KdV.l} h_1 h_2}{6}\frac{\rho_1 h_1 +]\rho_2 h_2}{\rho_2 h_1 + \rho_1 h_2}. \tag{4}$$

Here, g is the gravity acceleration set to 9.80 m s^{-2}. The thicknesses of the upper and lower layers (h_1 and h_2) were determined based on the depth of the maximum density gradient from the density profiles obtained from the UCTD. The densities at the upper and lower layers (ρ_1 and ρ_2, respectively) were determined as the minimum density within the upper layer and the maximum density within the lower layer, respectively. The solution of Equation (1) for the displacement $\eta(x,t)$ yields the nonlinear soliton as follows:

$$\eta(x,t) = \eta_0 \mathrm{sech}^2\left(\frac{x - c_{KdV.iw}t}{\Delta_{KdV.iw}}\right). \tag{5}$$

Here, the theoretical propagation speed $c_{KdV.iw}$ and characteristic width $2\Delta_{KdV.iw}$ were calculated from the linear phase speed $c_{KdV.\,l}$ and the amplitude (η_0) of the vertical displacement of η are as follows:

$$c_{KdV.iw} = c_{KdV.\,l} + \frac{|\alpha|\eta_0}{3}, \tag{6}$$

$$2\Delta_{KdV.iw} = 2\left(\frac{12\beta}{|\alpha|\eta_0}\right)^{1/2}. \tag{7}$$

By considering cubic nonlinearity, Equation (1) becomes as follows, yielding the extended KdV (eKdV) theory [51]:

$$\frac{\partial \eta}{\partial t} + c_{KdV.l}\frac{\partial \eta}{\partial x} + \alpha\eta\frac{\partial \eta}{\partial x} + \alpha_1\eta^2\frac{\partial \eta}{\partial x} + \beta\frac{\partial^3 \eta}{\partial x^3} = 0. \tag{8}$$

Here, $\alpha_1 = \frac{3c_{KdV.l}}{(h_1 h_2)^2}\left[\frac{7}{8}\left(\frac{\rho_2 {h_1}^2 - \rho_1 {h_2}^2}{\rho_2 h_1 + \rho_1 h_2}\right)^2 - \left(\frac{\rho_2 {h_1}^3 + \rho_1 {h_2}^3}{\rho_2 h_1 + \rho_1 h_2}\right)\right]$ is a cubic nonlinear parameter in the two-layer system. The theoretical propagation speed $c_{eKdV.iw}$ and characteristic width $2\Delta_{eKdV.iw}$ based on the eKdV theory in the two-layered system are as follows:

$$c_{eKdV.iw} = c_{KdV.\,l} + \frac{|\alpha|\eta_0}{3} + \frac{\alpha_1 {\eta_0}^2}{6}, \tag{9}$$

$$2\Delta_{eKdV.iw} = 2\left(\frac{12\beta}{|\alpha|\eta_0 + 0.5\alpha_1{\eta_0}^2}\right)^{1/2}. \tag{10}$$

2.2.2. *Doppler Shift Method*

To estimate the propagation direction of NLIWs using the Doppler shift caused by propagating NLIWs observed from a moving ship, the theoretical propagation speed $c_{KdV.iw}$ and ship speed v_{sh} were assumed to be constant during the measurement period, and the propagation direction was assumed to be orthogonal to the constant phase lines parallel to the wavefront lines (Figure 2a,b). As the estimated propagation direction ϕ_{ds} is in reference to the ship course ϕ_{sh}, the apparent propagation speed c_{ap} can be represented as the difference between $c_{KdV.iw}$ and the ship speed in direction θ_{ds} as follows:

$$c_{ap} = c_{KdV.iw} - v_{sh}\cos(\theta_{ds}). \tag{11}$$

Here, $\theta_{ds} = |\phi_{sh} - \phi_{ds}|$ is the angular difference between the ship course and the propagation direction of the NLIWs. Because the Doppler-shifted apparent frequency f_{ap} or the inverse of the apparent period T_{ap} can be represented by c_{ap} and $c_{iw} = \lambda_{KdV.iw} f_{iw}$, where $\lambda_{KdV.iw}$ is the wavelength of the NLIWs and the Doppler equation $f_{ap} = f_{iw}\frac{c_{ap}}{c_{iw}}$ [52], the following equation can be used to estimate the ϕ_{ds}:

$$\frac{1}{T_{ap}} = f_{ap} = f_{iw}\frac{c_{ap}}{c_{iw}} = \frac{c_{ap}}{\lambda_{KdV.iw}}. \tag{12}$$

Here, T_{ap} is determined from measurements (Table 1), while $\lambda_{KdV.iw}$ is determined by the Cnoidal model [50] as

$$\lambda_{KdV.iw} = 2\Delta_{KdV.iw}K(s) \approx 3.7\Delta_{KdV.iw}, \tag{13}$$

where $K(s)$ is a complete elliptic integral of the first kind and parameter s is set to 0.5. Equation (11) can then be rewritten using Equation (12) as follows:

$$c_{KdV.iw} - v_{sh}\cos|\phi_{sh} - \phi_{ds}| = f_{ap}\lambda_{KdV.iw}. \tag{14}$$

Further, the ϕ_{ds} is obtained as follows:

$$\phi_{ds} = \phi_{sh} \pm \cos^{-1}\left(\frac{c_{KdV.iw} - f_{ap}\lambda_{KdV.iw}}{v_{sh}}\right). \tag{15}$$

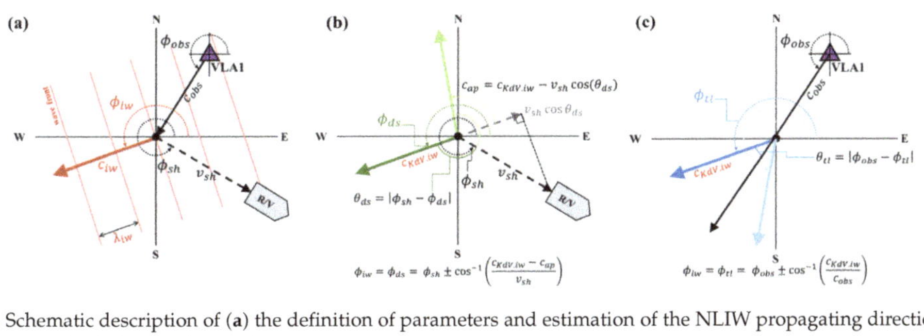

Figure 2. Schematic description of (**a**) the definition of parameters and estimation of the NLIW propagating direction based on the (**b**) Doppler shift and (**c**) time lag.

The propagation direction of the NLIWs estimated using the method described above has an angular ambiguity caused by the sign of the arccosine part. Thus, a physically reasonable direction between the two was selected. The ϕ_{ds} and ϕ_{sh} are angles in degrees measured counter-clockwise from the east (for example, 180° and 270° correspond to the westward and southward directions, respectively).

2.2.3. *Time Lag Method*

Independent of the method described in Section 2.2.2, the propagation direction of NLIWs was estimated using the distance between two locations of the NLIW front observed at different times, with the assumption that the NLIWs propagate across the two measurement locations with an angle orthogonal to the constant phase lines at a constant speed (Figure 2a,c). The observed propagation speed was estimated by dividing the distance between the two locations D_{obs} by the arrival time lag T_{obs}, for example, $c_{obs} = \frac{D_{obs}}{T_{obs}}$. Then, $c_{KdV.iw}$ was calculated from c_{obs} and angular difference $\theta_{tl} = |\phi_{obs} - \phi_{tl}|$ between ϕ_{obs} (direction from the first measurement location to the second measurement location) and the propagation direction of NLIWs ϕ_{tl} as

$$c_{KdV.iw} = c_{obs}\cos(\theta_{tl}) = \frac{D_{obs}}{T_{obs}}\cos(|\phi_{obs} - \phi_{tl}|). \tag{16}$$

Finally, the ϕ_{tl} was obtained from Equation (16) as

$$\phi_{tl} = \phi_{obs} \pm \cos^{-1}\left(\frac{c_{KdV.iw}T_{obs}}{D_{obs}}\right). \tag{17}$$

The propagation direction estimated using the method described above (*time lag* method) also has an angular ambiguity caused by the sign of the arccosine part. Thus, a physically reasonable direction is selected. The ϕ_{tl} and ϕ_{obs} are angles in degrees measured counter-clockwise from the east.

2.2.4. Estimation of Propagation Speed and Direction

Two propagation directions of NLIWs estimated from the two methods were used to estimate the optimal propagation direction and successive propagation speed. First, a consistent direction between the two directions derived from each method was selected to minimize the ambiguity where the two methods yield angular difference $|\phi_{ds} - \phi_{tl}|$ typically less than 30°. For example, each one (bold green and blue colors in Figure 2b,c) among two ϕ_{ds} and two ϕ_{tl} are selected as more consistent between the two methods and physically reasonable, while inconsistent ϕ_{ds} and ϕ_{tl} (deemed green and blue colors in Figure 2b,c) among the two ϕ_{ds} and two ϕ_{tl} were not selected. To optimize the propagation speed and direction, the difference in consistent propagation directions from the two methods $|\phi_{ds} - \phi_{tl}|$ was minimized by iteratively changing $c_{KdV.iw}$ at intervals of 0.01 m·s^{-1} instead of using the constant propagation speed derived from the two-layered KdV theory. The updated propagation speed was determined from the iterations, and the final propagation direction of NLIWs ϕ_{iw} was determined by averaging the two directions $(\phi_{ds} + \phi_{tl})/2$ when $|\phi_{ds} - \phi_{tl}|$ reached its minimum value for the updated propagation speed. Iterations were performed for a range of 30% deviation from $c_{KdV.iw}$ (typically requiring 38 iterations to reach the minimum), which is comparable to the range of the interannual variation of propagation speed reported in a previous study [47], as discussed in Section 4.

2.2.5. Estimation of Propagation Direction Using Satellite Images

The propagation directions from the MODIS images (Figure 3) were estimated from the horizontal curvature of the leading fronts from the sea surface manifestations [28]. The propagation direction was calculated by the direction of the center of the straight line connecting the endpoints of the manifestations to the center of the arc for the manifestations. For example, a manifestation of NLIW is shown in the blue box in Figure 3a,c. The orange curvature line from both endpoints (Points A and B) is the leading front of the NLIW. Point C is the center of line AB. Point D is the center of the arc AB. In this case, the angle between the vectors from C to D, measured counter-clockwise from the east, is the propagation direction.

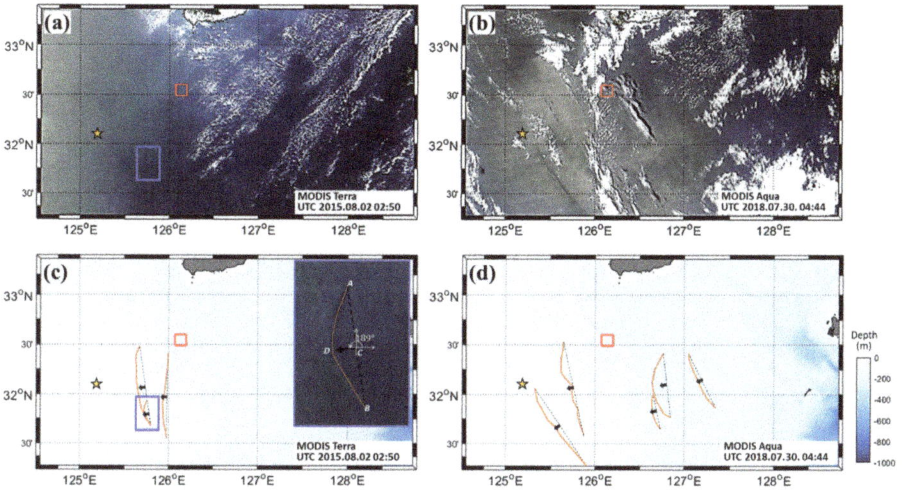

Figure 3. (**a**,**b**) Two satellite images where (**c**,**d**) surface manifestations of NLIWs (orange lines) are confirmed, from MODIS Terra at 02:50 UTC on 02 August 2015, and MODIS Aqua at 04:44 UTC on 30 July 2018. The blue box is the example of the propagation direction calculation. The SAVEX15 and IORS are marked by a red box and yellow star, respectively.

3. Applications

3.1. SAVEX15

On 23 May 2015 during the SAVEX15, the existence of NLIWs was confirmed from a series of UCTD profiling measurements, for example, at 15:56 UTC (vertical red dashed line in Figure 4b), and the time-depth pattern of temperature variations observed at the northern mooring VLA1 at 15:02 UTC (vertical grey dashed line in Figure 4a). The NLIW was observed at VLA1 3239 s earlier than that observed in the UCTD measurements (Figures 4 and 5a). The 13 °C isotherm displacements, derived from the UCTD and VLA1 observations, commonly range from 27 to 35 m (Figure 5a). The NLIW has an amplitude (η_0) of 6.1 and 5.9 m estimated from UCTD and VLA1 observations, respectively (Figures 4 and 5a). The two-layered theoretical parameters of the NLIWs are listed in Table 1, along with the estimated parameters, where the upper and lower layer densities (ρ_1 and ρ_2) are 1023.35 and 1025.17 kg·m^{-3}, respectively, and the upper and lower layer thicknesses (h_1 and h_2) are 34.8 and 65.2 m, respectively (Figure 5b and Table 1).

Table 1. Nonlinear internal wave parameters for the two cases observed, with calculated and estimated values.

	Parameters	SAVEX15	IORS18
Observation			
ρ_1 (kg·m^{-3})	Upper layer density	1023.35	1018.18
ρ_2 (kg·m^{-3})	Lower layer density	1025.17	1024.99
h_1 (m)	Upper layer thickness	34.8	24.0
h_2 (m)	Lower layer thickness	65.2	28.0
η_0 (m)	Wave amplitude	6.1	7.0
T_{ap} (sec)	Wave apparent period	1695	237
	Ship course	148	334

Table 1. *Cont.*

	Parameters	SAVEX15	IORS18
v_{sh} $(\mathrm{m{\cdot}s^{-1}})$	Ship speed	0.47	1.38
T_{obs} (sec)	Time lag between two different observations	3239	2370
D_{obs} (m)	Distance between two different observations	2233	3045
C_{obs} $(\mathrm{m{\cdot}s^{-1}})$	Speed between two different observations	0.69	1.29
ϕ_{obs} (°)	Angle between two different observations	230	243
Estimation base on KdV and eKdV theory			
α $(\mathrm{s^{-1}})$	Nonlinear parameter	−0.0127	−0.0041
α_1 $(\mathrm{s^{-1}{\cdot}m^{-1}})$	Cubic nonlinear parameter	−0.0009	−0.0011
β $(\mathrm{m^3{\cdot}s^{-1}})$	Dispersion parameter	238	104
$c_{KdV.l}$ $(\mathrm{m{\cdot}s^{-1}})$	Linear phase speed	0.63	0.92
$c_{KdV.iw}$ $(\mathrm{m{\cdot}s^{-1}})$	Propagating speed based on KdV theory	0.65	0.93
$c_{eKdV.iw}$ $(\mathrm{m{\cdot}s^{-1}})$	Propagating speed based on eKdV theory	0.65	0.92
$\Delta_{KdV.iw}$ (m)	Characteristic width based on KdV theory	384	416
$\Delta_{eKdV.iw}$ (m)	Characteristic width based on eKdV theory	432	1182
$\lambda_{KdV.iw}$ (m)	Wavelength based on KdV theory	712	772
$\lambda_{eKdV.iw}$ (m)	Wavelength based on eKdV theory	802	2192
Estimation based on proposed methods (KdV theory)			
θ_{ds} (°)	Angle between ϕ_{ds} and ϕ_{sh}	60	126
ϕ_{ds} (°)	NLIW propagating direction from Doppler shift	208	208
θ_{tl} (°)	Angle between ϕ_{tl} and ϕ_{obs}	19	45
ϕ_{tl} (°)	NLIW propagating direction from time lag	211	198
c_{iw} (°)	Optimal NLIW propagating speed	0.64	1.06
ϕ_{iw} (°)	Optimal NLIW propagating direction	210	205
Estimation based on proposed methods (eKdV theory)			
θ_{ds} (°)	Angle between ϕ_{ds} and ϕ_{sh}	68	100
ϕ_{ds} (°)	NLIW propagating direction from Doppler shift	215	235
θ_{tl} (°)	Angle between ϕ_{tl} and ϕ_{obs}	18	46
ϕ_{tl} (°)	NLIW propagating direction from time lag	212	197
c_{iw} (°)	Optimal NLIW propagating speed	0.66	1.30
ϕ_{iw} (°)	Optimal NLIW propagating direction	213	234

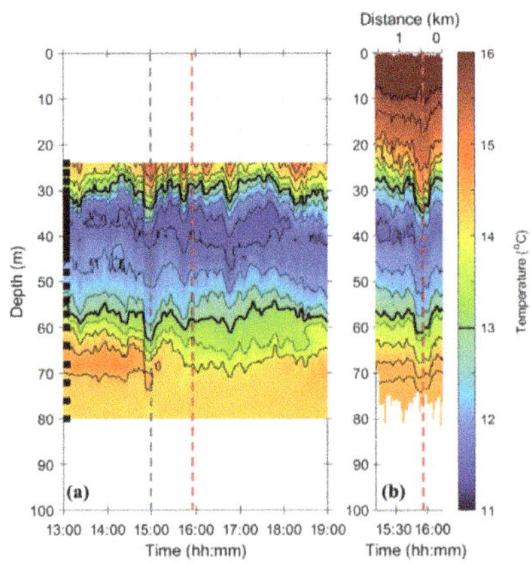

Figure 4. Time-depth pattern of water temperature observed from (**a**) northern mooring (VLA1) and (**b**) UCTD during the SAVEX15. The contour interval is 0.5 °C. The 13 °C isotherm is denoted by a thick black line. The times of NLIW observations at the VLA1 and UCTD are denoted by vertical grey and red dashed lines, respectively. Depths of thermistors attached to the VLA1 are denoted by black squares on the left axes.

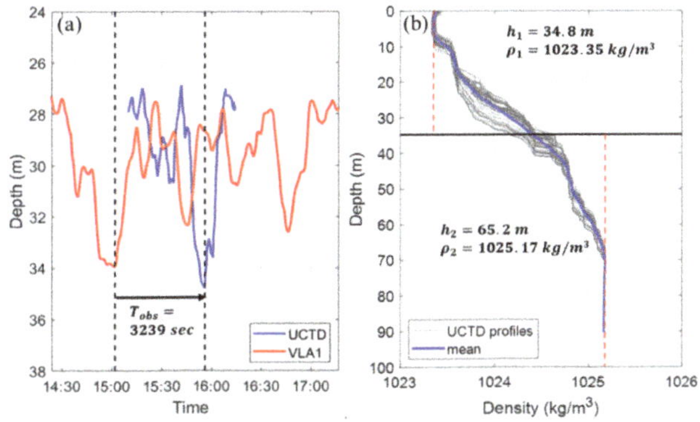

Figure 5. (**a**) Time-series of 13 °C isotherm depth observed at the VLA1 (red) and UCTD (blue), and (**b**) vertical profiles of density obtained from the UCTD measurements (grey) at 15:10–16:14 UTC on 23 May 2015. In (**b**), the average profile is marked in blue, and minimum and maximum densities at the upper and lower layers (corresponding to ρ_1 and ρ_2), respectively, are shown with red dashed lines.

When the ship moved at a speed of 0.47 m·s^{-1} and a direction of 148° (northwestward), the apparent propagation direction ϕ_{ds} of NILWs had an angular difference θ_{ds} of $\pm 60°$ with ship course ϕ_{sh} (Figure 6a), derived from the *Doppler shift* method using Equation (15), resulting in]$\phi_{ds} = 208°$ (southwestward) or $\phi_{ds} = 88°$ (northward). From the distance (D_{obs} = 2233 m) between the two measurement locations (VLA1 and ship) and the time lag of the NLIW arrivals (T_{obs} = 3239 s), the observed propagation direction of NILWs ϕ_{tl} was estimated to have an angular difference θ_{tl} of $\pm 19°$ with ϕ_{obs} (Figure 6b), derived from the *time lag* method using Equation (17), resulting in $\phi_{tl} = 211°$ (southwestward) or $\phi_{tl} = 249°$ (slightly more southwestward). Thus, more consistent propagation directions of $\phi_{ds} = 208°$ and $\phi_{tl} = 211°$ were selected to optimize the propagation speed and direction. By minimizing $|\phi_{tl} - \phi_{ds}|$, the optimal propagation speed (c_{iw}) of 0.64 m·s^{-1} was derived from the iterative calculations, yielding $\phi_{ds} = 208°$ and $\phi_{tl} = 211°$ with $|\phi_{tl} - \phi_{ds}| = 3°$, and the resultant propagation direction (ϕ_{iw}) of 210° (southwestward).

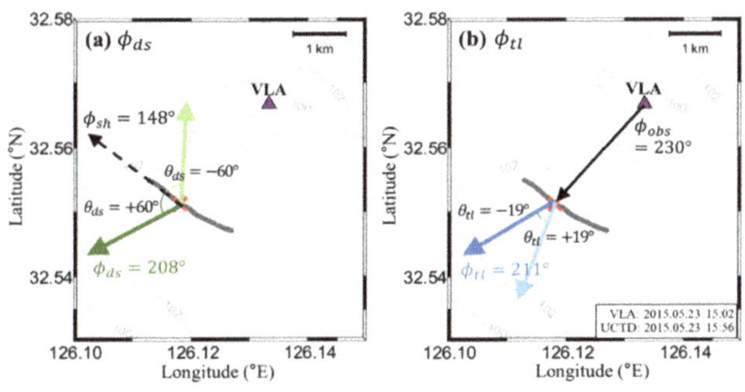

Figure 6. Propagation direction of NLIW observed during the SAVEX15, estimated from (**a**) *Doppler shift* method and (**b**) *time lag* method. The ϕ_{sh}, ϕ_{obs}, θ_{ds}, and θ_{tl} are labelled in the plots. Date and times of the corresponding events are noted in the right bottom corner.

3.2. IORS18

On 30 August 2018 during the IORS18, the existence of NLIWs was confirmed from the isotherm displacements observed by a series of vertical profiling UCTD measurements, particularly at 11:11 UTC (vertical red dashed line in Figure 7b), and the time-depth pattern of temperature measurements at the IORS, at 10:32 UTC (vertical grey dashed line in

Figure 7a). The NLIWs observed at IORS were 2370 s earlier than those observed at the UCTD (Figure 8a). The 18 °C isotherm displacements observed from the UCTD and IORS were comparable, ranging from 23 to 31 m (Figure 8a). The NLIWs (the leading NLIW among a set observed by the UCTD) had an amplitude (η_0) of 6.8 and 7.0 m derived from the IORS and UCTD observations, respectively (Figures 7 and 8a). The densities at the upper and lower layers (ρ_1 and ρ_2) were 1018.18 and 1024.99 kg·m^{-3}, and thicknesses of the upper and lower layers (h_1 and h_2) were 24.0 and 28.0 m, respectively (Figure 8b and Table 1).

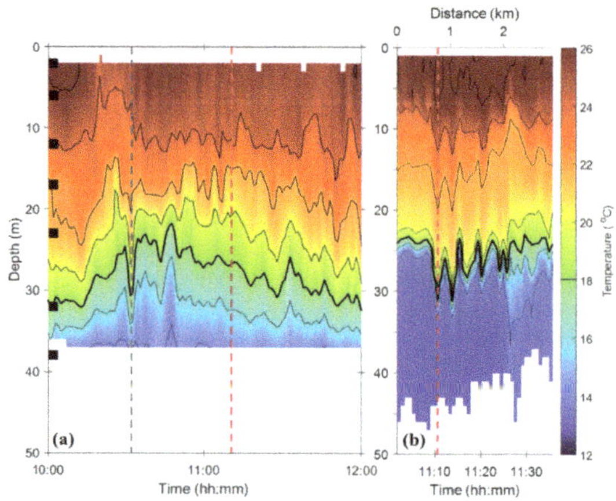

Figure 7. Time-depth pattern of water temperature observed at the (**a**) IORS and (**b**) UCTD during IORS18. The contour interval is 2 °C. The 18 °C isotherm is denoted by a thick black line. Time of NLIW observations at the IORS and UCTD are denoted by vertical grey and red dashed lines, respectively. Depths of thermistors attached to the IORS are denoted by black squares on the left axes.

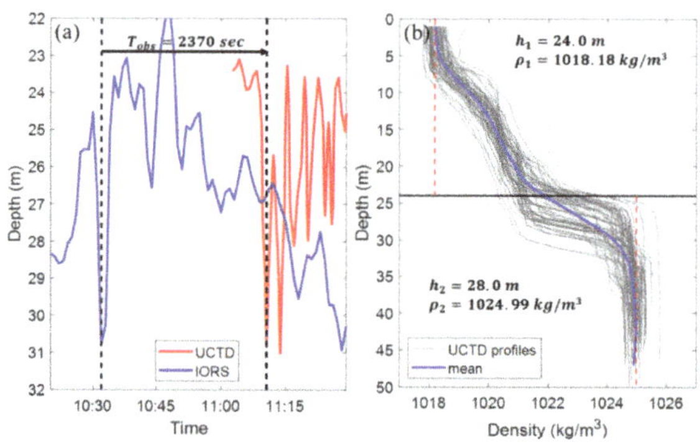

Figure 8. Information related to NLIWs observed at the IORS and UCTD during IORS18. (**a**) Time-series of the depth of the 18 °C isotherm observed at the IORS (red) and UCTD (blue), and (**b**) vertical profiles of density obtained from the UCTD measurements (grey) at 11:02–11:59 UTC on 30 August 2018. In (**b**), the average profile is marked in blue, and minimum and maximum densities at the upper and lower layers (corresponding to ρ_1 and ρ_2), respectively, are shown with red dashed lines.

When the ship moved at a speed of 1.38 m·s^{-1} and a direction of 334° (southeastward), the apparent propagation direction ϕ_{ds} of NLIWs had the angular difference θ_{ds} of ±126° with the ship course ϕ_{sh} (Figure 9a), derived from the *Doppler shift* method using Equation (15), resulting in $\phi_{ds} = 208°$ (southwestward) or $\phi_{ds} = 100°$ (northward). From the distance (D_{obs} = 3045 m) between the two measurement locations (IORS and ship) and the time lag of the NLIW arrivals (T_{obs} = 2370 s), the observed propagation direction of NLIW ϕ_{tl} was estimated to have an angular difference $\theta_{tl} = \pm 45^{\circ}$ with ϕ_{obs} (Figure 9b), derived from the *time lag* method using Equation (17), resulting in $\phi_{tl} = 198^{\circ}$ (southwestward) or $\phi_{tl} = 288^{\circ}$ (southeastward). Thus, more consistent propagation directions of $\phi_{ds} = 208^{\circ}$ and $\phi_{tl} = 198^{\circ}$ were selected to optimize the propagation speed and direction. By minimizing $||\phi_{tl} - \phi_{ds}|$, optimal propagation speed (c_{iw}) of 1.06 m·s^{-1} was derived from the iterative calculations, yielding $\phi_{ds} = 205^{\circ}$ and $\phi_{tl} = 205^{\circ}$ with $|\phi_{tl} - \phi_{ds}| = 0^{\circ}$, and the resulting propagation direction (ϕ_{iw}) of 205° was obtained.

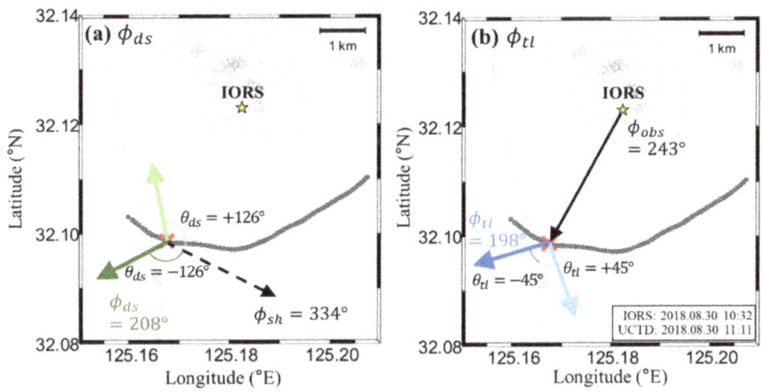

Figure 9. Propagation direction of NLIW observed during IORS18, estimated from (**a**) *Doppler shift* method and (**b**) *time lag* method. The ϕ_{sh}, ϕ_{obs}, θ_{ds}, and θ_{tl} are labeled in the plots. Dates and times of the corresponding events are noted in the right bottom corner.

4. Discussion

Herein, we discussed whether the propagation speeds of NLIWs estimated using the proposed method are reasonable based on the KdV theory and previous observations. Interannual variations of the theoretical propagation speed ($c_{KdV.iw}$, NIFS-SAVEX15) in May from 1994 to 2019 derived from the NIFS historical hydrographic data near the SAVEX15 area range from 0.36 to 0.71 m·s^{-1}, with a temporal mean and standard deviation of 0.50 and 0.09 m·s^{-1}, respectively (red line in Figure 10a). A long-term decreasing trend was observed in May $c_{KdV.iw}$ (NIFS-SAVEX15) at a rate of −0.004 m·s^{-1}·yr^{-1} (red dotted line in Figure 10a), primarily because of the decreasing density stratification, that is, increasing ρ_1 and decreasing ρ_2 with no significant change in h_1 and h_2 in May (red lines in Figure 10b,c). The propagation speed for May 2015 estimated using the proposed method (c_{iw}) was consistent with the theoretical propagation speed ($c_{KdV.iw}$; SAVEX15) derived from the hydrographic data obtained during SAVEX15 and that (NIFS-SAVEX15) derived from the nearby NIFS data, with minor (<0.05 m·s^{-1}) differences (closed square, open square, and open circle in Figure 10a).

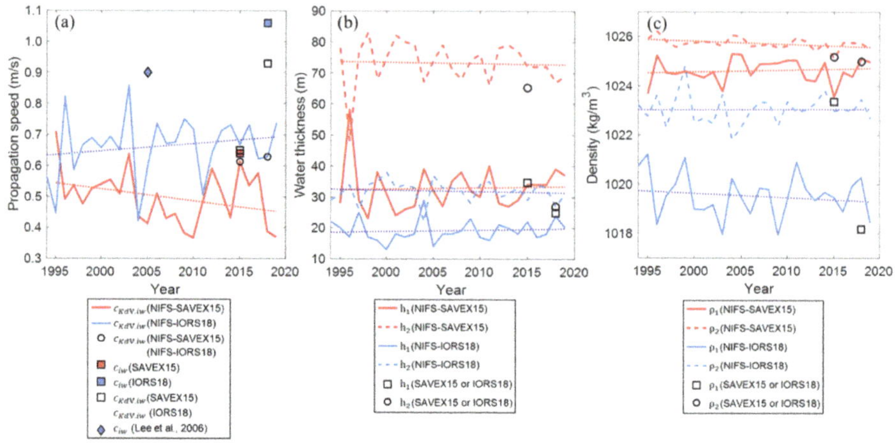

Figure 10. (**a**) Time series of propagation speed estimated from the method proposed in this study (colored squares; red for SAVEX15 and blue for IORS18) and KdV theory (open squares, open circles, and colored lines). KdV theory was applied using the NIFS historical hydrographic data for May (red line and open circle) and August (blue line and open circle), and hydrographic data obtained during SAVEX15 and IORS18 (open squares). The propagation speed of NLIWs observed in the northern ECS in August 2005 [33] is denoted by a blue diamond. Time series of (**b**) thickness (h_1 and h_2) and (**c**) density (ρ_1 and ρ_2) of the upper (solid lines) and lower (dashed lines) layers derived from the NIFS historical hydrographic data for May (red) and August (blue), and hydrographic data obtained during SAVEX15 and IORS18 (open squares and circles for upper and lower layers, respectively). Long-term trends of (**a**) propagation speed, (**b**) layer thickness, and (**c**) layer density from 1994 to 2019 are remarked with dotted lines.

Similarly, interannual variations in the theoretical propagation speed ($c_{KdV.iw}$, NIFS-IORS18) in August from 1994 to 2019 derived from the NIFS historical hydrographic data near the IORS ranged from 0.42 to 0.86 m·s^{-1}, with a temporal mean and standard deviation of 0.66 and 0.10 m·s^{-1}, respectively (blue line in Figure 10a). However, the long-term trend in August $c_{KdV.iw}$ (NIFS-IORS18) was positive (at a rate of 0.003 m·s^{-1}·yr^{-1}; blue dotted line in Figure 10a) because of the increasing density stratification between the layers, that is, decreasing ρ_1 and increasing ρ_2 with no significant change in h_1 and h_2 in August (blue lines in Figure 10b,c). In contrast to May 2015, the propagation speed for August 2018 estimated using the proposed method (c_{iw}) was not consistent with the theoretical propagation speed ($c_{KdV.iw}$; IORS18) derived from the hydrographic data obtained during IORS18 and that (NIFS-IORS18) derived from the nearby NIFS data, yielding significant (>0.25 m·s^{-1}) differences (closed square, open square, and open circle in Figure 10a). The difference in density stratification between the NIFS and IORS18 data cannot explain the difference in c_{iw} from the theoretical propagation speeds of $c_{KdV.iw}$ (IORS18) and $c_{KdV.iw}$ (NIFS-IORS18), implying the limitation of theoretical estimation. A similar difference in the observed propagation speed from the theoretical propagation speed in the area near the IORS was reported from the observations in August 2005 (blue diamond in Figure 10) [33].

To determine whether the propagation directions of NLIWs estimated using the proposed method are physically reasonable, we compared the ϕ_{iw} values with those derived from satellite images and previous observations. The ϕ_{iw} values estimated during the two experiments (SAVEX15 and IORS18) using the proposed method were consistent with those derived from MODIS images. Despite the fact that the surface manifestations of NLIWs observed in the two MODIS images were distant (56–123 km from the SAVEX15 area and 42–190 km from the IORS) from the locations of NLIWs observed during the two experiments, south-westward-propagating NLIWs (propagation direction of 186–209°) were consistently found in the two images (Figure 3). Previous observations based on satellite SAR and optical images taken between 1993 and 2004 [35] and SAR images taken between 2014 and 2015 [36] in the northern ECS also support the south-westward-propagating NILWs (propagation direction ranging from 212° to 245°), which may be dominant among the various NLIWs propagating in multiple directions from multiple sources in the northern ECS (Figure 1a).

The two-layered classical (ordinary) KdV theory used in this study has clear limitations. The classical KdV theory is very simplified and assumes weak non-linearity and weak dispersiveness. In fact, NLIWs observed in many areas have been better explained by the eKdV theory than by the KdV theory. However, propagation speeds, in cases of SAVEX15 and IORS18, derived based on the eKdV theory ($c_{eKdV.iw}$) including the cubic non-linearity are not significantly different from those based on the KdV theory ($c_{KdV.iw}$), yielding the difference less than 0.02 m·s^{-1} due to relatively small η_0 (Table 1). Furthermore, as in the case of NLIWs in the South China Sea, finite-depth theory may be theoretically more appropriate than shallow-water theory where the KdV and eKdV theories are based on [53]. The theoretical propagation speeds in the forms of Equations (6) and (9) are limited to the case of no background pedestal condition that could not be considered in this study and might affect the speed significantly. The rigid lid assumption of the KdV and eKdV theories at the top boundary is not fully realistic, although reasonable in many cases, because the resonant interaction between the surface and internal waves supports the possible need for the presence of a free surface at the top boundary to yield more realistic theoretical estimates [54]. In addition, the results presented in this study are limited to only mode-1 NLIWs by applying the two-layered system, yet the vertical profiles of mean density observed during the two experiments (Figures 5b and 8b) support normal mode decompositions (J. Klinck's Matlab program dynmodes.m, available online at http://github.com/sea-mat/dynmodes; accessed on 5 October 2021) for the first three modes corresponding to 49%, 18%, and 14% for SAVEX15 and 50%, 21%, and 13% for IORS18, respectively. Multi-mode NLIWs beyond the mode-1 NLIWs in the region, not investigated in this study, yet explaining about half of NLIWs, need to be examined in the future.

5. Concluding Remarks

We present a novel method to estimate the propagation speed and direction of NLIWs using widely collected underway and moored observations, and the results of applying the method to two cases of NLIWs observed in May 2015 (SAVEX15) and August 2018 (IORS18). Two-layered KdV theory and satellite images were used to discuss the results of the proposed method. The propagation direction of NLIWs was estimated with respect to a moving ship using the Doppler shift relationship (1. *Doppler shift* method) and independently using the time lag between the NLIWs observed at two different locations (2. *time lag* method). Then, the propagation speed and direction were optimized to minimize the difference in propagation directions derived from the two methods by iterating the propagation speed in the range of ±30% at a resolution of 0.01 m·s^{-1}. The results derived from the proposed method are robust, as the range of iterative propagation speeds is comparable to the interannual variation of theoretical propagation speeds estimated using historical hydrographic data, yielding an error of less than 15% for the propagation direction. Because in situ observations of NLIWs are still challenging to collect and propagation speed and direction cannot be directly measured from subsurface instruments, our proposed method for estimating the propagation speed and direction of NLIWs using common underway and moored measurements is of practical importance, particularly over a broad shelf, such as the northern ECS, where the multi-directional propagation of multi-mode NLIWs from multiple sources is often observed.

Author Contributions: Conceptualization, S.-W.L. and S.-H.N.; methodology, S.-W.L. and S.-H.N.; validation, S.-W.L. and S.-H.N.; formal analysis, S.-W.L.; investigation, S.-W.L. and S.-H.N.; data curation, S.-W.L.; writing—original draft preparation, S.-W.L.; writing—review and editing, S.-W.L. and S.-H.N.; visualization, S.-W.L.; supervision, S.-H.N.; project administration, S.-H.N. Both authors have read and agreed to the published version of the manuscript.

Funding: SAVEX15 was funded by the US ONR (N00014-13-1-0510), KIOST (PE99531), and KRISO (PES1940). This work was funded by the Ministry of Oceans and Fisheries, Republic of Korea through the "Establishment of the Ocean Research Station in the Jurisdiction Zone and Convergence Research" and "Deep Water Circulation and Material Cycling in the East Sea (20160040)".

Institutional Review Board Statement: Not applicable.

Informed Consent Statement: Not applicable.

Data Availability Statement: All data supporting this manuscript are available from the corresponding author upon reasonable request. SAVEX15 data are available from Lee et al. [43], IORS18 data are available at the Korea Institute of Ocean Science and Technology (KIOST; https://kors.kiost.ac.kr/en/; accessed on 5 October 2021), historical CTD data are available at NIFS (http://www.nifs.go.kr/kodc/eng/index.kodc; accessed on 5 October 2021), and MODIS satellite images are available at NASA (http://worldview.earthdata.nasa.gov; accessed on 5 October 2021).

Acknowledgments: We would like to thank Heechun Song, Sea-Moon Kim, and Byoung-Nam Kim, all individuals who participated in the SAVEX15 experiment, and Jae-Seol Shim, Jin-Young Jeong, Yong Sun Kim, In-Ki Min, Kwang-Young Jeong, and Da-Eun Yeo who participated in IORS18 experiments and provided technical support for the IORS.

Conflicts of Interest: The authors declare no conflict of interest.

References

1. MacKinnon, J.A.; Gregg, M.C. Mixing on the late-summer new England shelf-Solibores, shear, and stratification. *J. Phys. Oceanogr.* **2003**, *33*, 1476–1492. [CrossRef]
2. Alford, M.H. Redistribution of energy available for ocean mixing by long-range propagation of internal waves. *Nature* **2003**, *423*, 159–162. [CrossRef] [PubMed]
3. D'Asaro, E.A.; Lien, R.-C.; Henyey, F. High-Frequency Internal Waves on the Oregon Continental Shelf. *J. Phys. Oceanogr.* **2007**, *37*, 1956–1967. [CrossRef]
4. Moum, J.N.; Farmer, D.M.; Shroyer, E.L.; Smyth, W.D.; Armi, L. Dissipative Losses in Nonlinear Internal Waves Propagating across the Continental Shelf. *J. Phys. Oceanogr.* **2007**, *37*, 1989–1995. [CrossRef]
5. Shroyer, E.L.; Moum, J.N.; Nash, J.D. Mode 2 waves on the continental shelf: Ephemeral components of the nonlinear internal wavefield. *J. Geophys. Res. Space Phys.* **2010**, *115*. [CrossRef]
6. Scotti, A.; Pineda, J. Plankton accumulation and transport in propagating nonlinear internal fronts. *J. Mar. Res.* **2007**, *65*, 117–145. [CrossRef]
7. Muacho, S.; da Silva, J.; Brotas, V.; Oliveira, P. Effect of internal waves on near-surface chlorophyll concentration and primary production in the Nazaré Canyon (west of the Iberian Peninsula). *Deep Sea Res. Part. I Oceanogr. Res. Pap.* **2013**, *81*, 89–96. [CrossRef]
8. Villamaña, M.; Mourino-Carballido, B.; Marañón, E.; Cermeno, P.; Chouciño, P.; da Silva, J.C.B.; Díaz, P.; Castro, B.F.; Gilcoto, M.; Graña, R.; et al. Role of internal waves on mixing, nutrient supply and phytoplankton community structure during spring and neap tides in the upwelling ecosystem of Ría de Vigo (NW Iberian Peninsula). *Limnol. Oceanogr.* **2017**, *62*, 1014–1030. [CrossRef]
9. Li, D.; Chou, W.-C.; Shih, Y.-Y.; Chen, G.-Y.; Chang, Y.; Chow, C.H.; Lin, T.-Y.; Hung, C.-C. Elevated particulate organic carbon export flux induced by internal waves in the oligotrophic northern South China Sea. *Sci. Rep.* **2018**, *8*, 2042. [CrossRef]
10. Drake, D.E.; Cacchione, D.A. Field observations of bed shear stress and sediment resuspension on continental shelves, Alaska and California. *Cont. Shelf Res.* **1986**, *6*, 415–429. [CrossRef]
11. Bogucki, D.; Dickey, T.; Redekopp, L.G. Sediment resuspension and transport by Sediment Resuspension and Mixing by Resonantly Generated Internal Solitary Waves solitary waves. *J. Phys. Oceanogr.* **1997**, *27*, 1181–1196. [CrossRef]
12. Klymak, J.M.; Moum, J.N. Internal solitary waves of elevation advancing on a shoaling shelf. *Geophys. Res. Lett.* **2003**, *30*. [CrossRef]
13. Lynch, J.; Newhall, A. Shallow-Water Acoustics. In *Applied Underwater Acoustics*; Elsevier: Amsterdam, The Netherlands, 2017; pp. 403–467. [CrossRef]
14. Apel, J.R.; Ostrovsky, L.; Stepanyants, Y.A.; Lynch, J.F. Internal solitons in the ocean and their effect on underwater sound. *J. Acoust. Soc. Am.* **2007**, *121*, 695–722. [CrossRef] [PubMed]
15. Boegman, L.; Stastna, M. Sediment Resuspension and Transport by Internal Solitary Waves. *Annu. Rev. Fluid Mech.* **2019**, *51*, 129–154. [CrossRef]
16. Lee, O.S. Observations on internal waves in shallow water. *Limnol. Oceanogr.* **1961**, *6*, 312–321. [CrossRef]
17. Ramp, S.; Tang, T.Y.; Duda, T.; Lynch, J.; Liu, A.; Chiu, C.-S.; Bahr, F.; Kim, H.-R.; Yang, Y.-J. Internal Solitons in the Northeastern South China Sea Part I: Sources and Deep Water Propagation. *IEEE J. Ocean. Eng.* **2004**, *29*, 1157–1181. [CrossRef]
18. Moum, J.N.; Nash, J.D. Seafloor Pressure Measurements of Nonlinear Internal Waves. *J. Phys. Oceanogr.* **2008**, *38*, 481–491. [CrossRef]

19. Shroyer, E.L.; Moum, J.N.; Nash, J. Observations of Polarity Reversal in Shoaling Nonlinear Internal Waves. *J. Phys. Oceanogr.* **2009**, *39*, 691–701. [CrossRef]
20. Alford, M.H.; Lien, R.-C.; Simmons, H.; Klymak, J.; Ramp, S.; Yang, Y.-J.; Tang, D.; Chang, M.-H. Speed and Evolution of Nonlinear Internal Waves Transiting the South China Sea. *J. Phys. Oceanogr.* **2010**, *40*, 1338–1355. [CrossRef]
21. Journal of Geophysical Research. *Nature* **1955**, *175*, 238. [CrossRef]
22. McSweeney, J.; Lerczak, J.A.; Barth, J.A.; Becherer, J.; Colosi, J.A.; MacKinnon, J.A.; MacMahan, J.H.; Moum, J.N.; Pierce, S.D.; Waterhouse, A.F. Observations of Shoaling Nonlinear Internal Bores across the Central California Inner Shelf. *J. Phys. Oceanogr.* **2020**, *50*, 111–132. [CrossRef]
23. Scotti, A.; Butman, B.; Beardsley, R.C.; Alexander, P.S.; Anderson, S. A Modified Beam-to-Earth Transformation to Measure Short-Wavelength Internal Waves with an Acoustic Doppler Current Profiler. *J. Atmos. Ocean. Technol.* **2005**, *22*, 583–591. [CrossRef]
24. Moum, J.N.; Smyth, W.D. The pressure disturbance of a nonlinear internal wave train. *J. Fluid Mech.* **2006**, *558*, 153–177. [CrossRef]
25. Mirshak, R.; Kelley, D.E. Inferring Propagation Direction of Nonlinear Internal Waves in a Vertically Sheared Background Flow. *J. Atmos. Ocean. Technol.* **2009**, *26*, 615–625. [CrossRef]
26. Chang, M.-H.; Lien, R.-C.; Yang, Y.J.; Tang, T.Y. Nonlinear Internal Wave Properties Estimated with Moored ADCP Measurements. *J. Atmos. Ocean. Technol.* **2011**, *28*, 802–815. [CrossRef]
27. Zhang, S.; Alford, M.H.; Mickett, J.B. Characteristics, generation and mass transport of nonlinear internal waves on the Washington continental shelf. *J. Geophys. Res. Oceans* **2015**, *120*, 741–758. [CrossRef]
28. Wang, J.; Huang, W.; Yang, J.; Zhang, H.; Zheng, G. Study of the propagation direction of the internal waves in the South China Sea using satellite images. *Acta Oceanol. Sin.* **2013**, *32*, 42–50. [CrossRef]
29. Greskowiak, J. Tide-induced salt-fingering flow during submarine groundwater discharge. *Geophys. Res. Lett.* **2014**, *41*, 6413–6419. [CrossRef]
30. Lindsey, D.T.; Nam, S.; Miller, S.D. Tracking oceanic nonlinear internal waves in the Indonesian seas from geostationary orbit. *Remote. Sens. Environ.* **2018**, *208*, 202–209. [CrossRef]
31. Karang, I.W.G.A.; Chonnaniyah; Osawa, T. Internal solitary wave observations in the Flores Sea using the Himawari-8 geostationary satellite. *Int. J. Remote. Sens.* **2019**, *41*, 5726–5742. [CrossRef]
32. Tensubam, C.M.; Raju, N.J.; Dash, M.K.; Barskar, H. Estimation of internal solitary wave propagation speed in the Andaman Sea using multi–satellite images. *Remote. Sens. Environ.* **2020**, *252*, 112123. [CrossRef]
33. Lee, J.H.; Lozovatsky, I.; Jang, S.-T.; Jang, C.J.; Hong, C.S.; Fernando, H.J.S. Episodes of nonlinear internal waves in the northern East China Sea. *Geophys. Res. Lett.* **2006**, *33*. [CrossRef]
34. Hsu, M.-K.; Liu, A.K.; Liu, C. A study of internal waves in the China Seas and Yellow Sea using SAR. *Cont. Shelf Res.* **2000**, *20*, 389–410. [CrossRef]
35. Alpers, W.; He, M.-X.; Zeng, K.; Guo, L.-F.; Li, X.-M. The distribution of internal waves in the East China Sea and the Yellow Sea studied by multi-sensor satellite images. In *Proceedings. 2005 IEEE International Geoscience and Remote Sensing Symposium, 2005. IGARSS'05*; IEEE: Piscataway, NJ, USA, 2005; Volume 7, pp. 4784–4787. [CrossRef]
36. Nam, S.; Kim, D.-J.; Lee, S.-W.; Kim, B.G.; Kang, K.-M.; Cho, Y.-K. Nonlinear internal wave spirals in the northern East China Sea. *Sci. Rep.* **2018**, *8*, 1–9. [CrossRef]
37. Liu, Z.; Wei, H.; Lozovatsky, I.; Fernando, H. Late summer stratification, internal waves, and turbulence in the Yellow Sea. *J. Mar. Syst.* **2009**, *77*, 459–472. [CrossRef]
38. Song, H.; Cho, C.; Hodgkiss, W.; Nam, S.; Kim, S.-M.; Kim, B.-N. Underwater sound channel in the northeastern East China Sea. *Ocean. Eng.* **2018**, *147*, 370–374. [CrossRef]
39. Park, J.; Seong, W.; Yang, H.; Nam, S.; Lee, S.-W. Array tilt effect induced by tidal currents in the northeastern East China Sea. *Ocean. Eng.* **2019**, *194*, 106654. [CrossRef]
40. Park, J.; Seong, W.; Yang, H.; Nam, S.; Lee, S.-W.; Choo, Y. Broadband acoustic signal variability induced by internal solitary waves and semidiurnal internal tides in the northeastern East China Sea. *J. Acoust. Soc. Am.* **2019**, *146*, 1110–1123. [CrossRef]
41. Byun, G.; Song, H.C. Adaptive array invariant. *J. Acoust. Soc. Am.* **2020**, *148*, 925–933. [CrossRef] [PubMed]
42. Yoon, S.; Yang, H.; Seong, W. Deep learning-based high-frequency source depth estimation using a single sensor. *J. Acoust. Soc. Am.* **2021**, *149*, 1454–1465. [CrossRef] [PubMed]
43. Lee, S.-W.; Nam, S.; Song, H. Moored and ship-based oceanographic data collected during Shallow-water Acoustic Variability EXperiment (SAVEX15). *SEANOE* **2020**. [CrossRef]
44. Ha, K.-J.; Nam, S.; Jeong, J.-Y.; Moon, I.-J.; Lee, M.; Yun, J.; Jang, C.J.; Kim, Y.S.; Byun, D.-S.; Heo, K.-Y.; et al. Observations Utilizing Korea Ocean Research Stations and their Applications for Process Studies. *Bull. Am. Meteorol. Soc.* **2019**, *100*, 2061–2075. [CrossRef]
45. Akima, H. A New Method of Interpolation and Smooth Curve Fitting Based on Local Procedures. *J. ACM* **1970**, *17*, 589–602. [CrossRef]
46. Ullman, D.S.; Hebert, D. Processing of Underway CTD Data. *J. Atmospheric Ocean. Technol.* **2014**, *31*, 984–998. [CrossRef]
47. Cho, C.; Nam, S.; Song, H. Seasonal variation of speed and width from kinematic parameters of mode-1 nonlinear internal waves in the northeastern East China Sea. *J. Geophys. Res. Oceans* **2016**, *121*, 5942–5958. [CrossRef]

48. Jackson, C. Internal wave detection using the Moderate Resolution Imaging Spectroradiometer (MODIS). *J. Geophys. Res. Space Phys.* **2007**, *112*. [CrossRef]
49. Korteweg, D.J.; de Vries, G. XLI: On the change of form of long waves advancing in a rectangular canal, and on a new type of long stationary waves. *Lond. Edinb. Dublin Philos. Mag. J. Sci.* **1895**, *39*, 422–443. [CrossRef]
50. Apel, J.R. A New Analytical Model for Internal Solitons in the Ocean. *J. Phys. Oceanogr.* **2003**, *33*, 2247–2269. [CrossRef]
51. Helfrich, K.R.; Melville, W.K. Long Nonlinear Internal Waves. *Annu. Rev. Fluid Mech.* **2006**, *38*, 395–425. [CrossRef]
52. Kinsler, L.E.; Frey, A.R.; Coppens, A.B.; Sanders, J.V. *Fundamentals of Acoustics*; John Wiley & Sons: Hoboken, NJ, USA, 1999.
53. Cai, S.; Xie, J.; Xu, J.; Wang, N.; Chen, Z.; Deng, X.; Long, X. Monthly variation of some parameters about internal solitary waves in the South China sea. *Deep Sea Res. Part I Oceanogr. Res. Pap.* **2013**, *84*, 73–85. [CrossRef]
54. Craig, W.; Guyenne, P.; Sulem, C. The surface signature of internal waves. *J. Fluid Mech.* **2012**, *710*, 277–303. [CrossRef]

Journal of
Marine Science and Engineering

Article

Nonseasonal Variations in Near-Inertial Kinetic Energy Observed Far below the Surface Mixed Layer in the Southwestern East Sea (Japan Sea)

Suyun Noh [1,2] and SungHyun Nam [1,3,*]

1 School of Earth and Environmental Science, College of Natural Science, Seoul National University, Seoul 08826, Korea; synoh17@snu.ac.kr
2 Future Innovation Institute, Siheung Campus, Seoul National University, Seoul 08826, Korea
3 Research Institute of Oceanography, College of Natural Science, Seoul National University, Seoul 08826, Korea
* Correspondence: namsh@snu.ac.kr; Tel.: +82-2-880-4138

Abstract: Near-inertial internal waves (NIWs) generated by surface wind forcing are intermittently enhanced below and within the surface mixed layer. The NIW kinetic energy below the surface mixed layer varies over intraseasonal, interannual, and decadal timescales; however, these variations remain unexplored, due to a lack of long-term, in situ observations. We present statistical results on the nonseasonal variability of the NIW kinetic energy 400 m below the surface mixed layer in the southwestern East Sea, using moored current measurements from 21 years. We used long time series of the near-inertial band (0.85–1.15 f) kinetic energy to define nine periods of relatively high (period high) and seven periods of relatively low (period low) NIW kinetic energy. The NIW kinetic energy average at period high was about 24 times higher than that at period low and those in specific years (2003, 2012–2013, 2016, and 2020) and decade (2010s) were significantly higher than those in other years and decade (2000s). Composite analysis revealed that negative relative vorticity and strong total strain significantly enhance NIW kinetic energy at 400 m. The relative vorticity was negative (total strain was positively enhanced) during seven (six) out of nine events of period high. NIW trapping in a region of negative relative vorticity and the wave capture process induce nonseasonal variations in NIW kinetic energy below the surface mixed layer. Our study reveals that, over intraseasonal, interannual, and decadal timescales, mesoscale flow fields significantly influence NIWs.

Keywords: near-inertial internal waves; nonseasonal variability; mesoscale flow field; relative vorticity; Okubo-Weiss parameter; subsurface mooring; southwestern East Sea; Japan Sea

Citation: Noh, S.; Nam, S. Nonseasonal Variations in Near-Inertial Kinetic Energy Observed Far below the Surface Mixed Layer in the Southwestern East Sea (Japan Sea). *J. Mar. Sci. Eng.* **2022**, *10*, 9. https://doi.org/10.3390/jmse10010009

Academic Editors: Eugenio Fraile-Nuez and Christos Stefanakos

Received: 13 November 2021
Accepted: 20 December 2021
Published: 23 December 2021

1. Introduction

Near-inertial internal gravity waves (NIWs), with frequencies close to the local inertial frequency (f), are ubiquitous in stratified rotating oceans. The NIWs mainly originate from surface wind forcing, enhanced primarily within the surface mixed layer; in general, the waves propagate into the ocean interior below the mixed layer and towards the equator and ultimately dissipate, while enhancing turbulent mixing [1–4]. The rate of work done by surface wind on global mixed layer NIWs is known to range from 0.3 TW to 1.3 TW, which is comparable to the global energy derived from barotropic to baroclinic tides (1.0–1.2 TW) [5–7]. Both winds and tides play a key role in providing energy to induce turbulent mixing and redistribute energy and materials in the ocean [3,8–13]. Along with turbulent mixing enhanced by tides, NIW-enhanced mixing may sustain the meridional overturning circulation [13–15]; previous studies used numerical simulations to suggest that near-inertial variations in the meridional overturning circulation are caused by equatorward-propagating NIWs [16,17]. NIWs are also important as they significantly affect, via turbulent fluxes, primary production and marine ecosystems [18–23].

The generation, evolution, propagation, and decay of NIWs are affected by mesoscale flow fields and wind forcing [2,24–27]. The wind forcing excites NIWs in the mixed layer which generally propagate equatorward horizontally and downward vertically below the mixed layer. The amount of wind energy input into and below the mixed layer is modulated by interaction processes between the mesoscale flow fields and NIWs. One method of interaction is the trapping (reflection) of NIWs in a region of negative (positive) relative vorticity that decreases (increases) the effective Coriolis frequency (e.g., $f_{eff} = f + \frac{1}{2}\zeta$ where ζ is the relative vorticity; $\zeta = \partial V/\partial x - \partial U/\partial y$, and U and V are zonal and meridional components of horizontal mesoscale currents) acting as a waveguide in the northern hemisphere (opposite sign in the southern hemisphere). Although, in general, NIWs freely propagate in frequencies between f and buoyancy frequency (N), the relative vorticity shifts the lowest limit from f to f_{eff} (Figure 1) [24,28,29]. Thus, the NIWs entering into a region of negative relative vorticity ($\zeta < 0$) or lowered f_{eff} can hardly propagate out of the region and can, thus, be trapped. Another way of interaction is a straining that stretches and rotates the wavenumber vectors of NIWs depending on differential advection of mesoscale flow fields (U, V). Under one kind of the straining processes called 'wave capture' (strain dominates vorticity), the NIWs can draw energy from mesoscale flow fields [30–33]. Staining causes exponential increase in the vertical and horizontal wavenumbers of NIWs, which results in growing wavenumbers and decreasing group velocities of NIWs, with NIWs captured within the region of high total strain and eventually dissipated through energy cascading (Figure 1) [30–33].

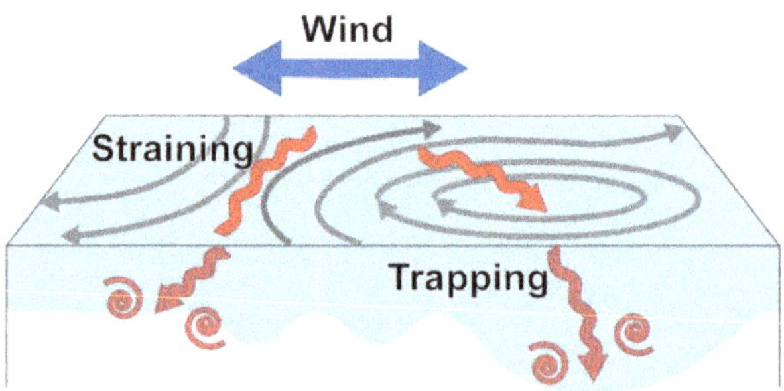

Figure 1. Schematics on interaction between mesoscale flow field (grey line) and wind-induced NIWs (red arrow), enhancing the near-inertial kinetic energy. Red spirals indicate dissipation of NIWs.

Previous studies on NIWs interacting with mesoscale flow fields in the East Sea (Sea of Japan) have been reported from both observations and numerical models. The formation, presence, and decay of mesoscale eddies are frequently observed in the East Sea, particularly off the east coast of Korea, and are partly associated with the strong meandering of a western boundary current, the East Korea Warm Current. A semi-permanent anticyclonic eddy, the Ulleung Warm Eddy (UWE), is often found off the coast where the boundary current forms and separates [34,35]. Seasonal variation in NIWs (with winter intensification) was reported by Mori et al. [36] and Jeon et al. [37] in association with the East Asian Monsoon and mesoscale circulation. Additionally, studies have suggested annual variations in the deep NIW kinetic energy observed off the coast, related to mesoscale fields imposed by the UWE [38]. Upward reflection of downward-propagating NIWs by the UWE was also observed [39]. Noh and Nam [40] reported the importance of mesoscale strain fields in the enhancement of NIWs, focusing on specific cases of NIW events. Likewise, the effects of mesoscale circulation on the behaviour of NIWs off the coast have been examined previously. However, the effects of mesoscale flow fields on temporal variations in NIW kinetic energy beyond the seasonal cycle have not been presented in the region to date.

Thus, this study is the first to describe the intraseasonal, interannual, and decadal variations (nonseasonal variations) in NIW kinetic energy in southwestern East Sea, using moored observations over the duration of 21 years. The objective was to identify statistically significant factors, particularly in association with mesoscale flow fields, that control nonseasonal variations in NIW kinetic energy based on long-term continuous observations. The data used and the methods applied in this study are described in the next section. In Section 3, we have presented the results of moored observations and damped slab model. Additionally, results of nonseasonal variations in NIW kinetic energy, in terms of surface wind forcing and mesoscale field variability, are provided and discussed in Section 4, and Section 5 provides the conclusions of our study.

2. Data and Methods

2.1. Data

Since 1996, long time-series data of zonal and meridional currents were collected using a subsurface mooring, named EC1 (37°19.13 N, 131°25.62 E), located between Ulleugdo and Dokdo, at a water depth of 2300 m (Figure 2). The EC1 was recovered and redeployed 24 times (as of December 2021) and equipped for most periods, with current meters at three nominal depths (400, 1400, and 2200 m). Rotary-type current meters (Aanderaa RCMs 7 and 8) and Doppler-type current meters (Aanderaa RCMs 9 and 11; Nortek Aquadopp) were attached to the mooring, and continuous time-series data were recorded with a sampling interval equal to or less than 1 h. An upward-looking acoustic Doppler current profiler (ADCP, 300 kHz) was mounted at 500 m with a depth interval (bin size) of 8 m, instead of a using single-type current meter at 400 m, from March 2011 to July 2012. All EC1 data collected from 1996 to 2020 were upgraded by quality control and quality assurance and were made available by SEANOE [41]. In this study, the time-series data of currents collected at 400 m of EC1 for almost 21 years, from January 2000 to November 2020, were used.

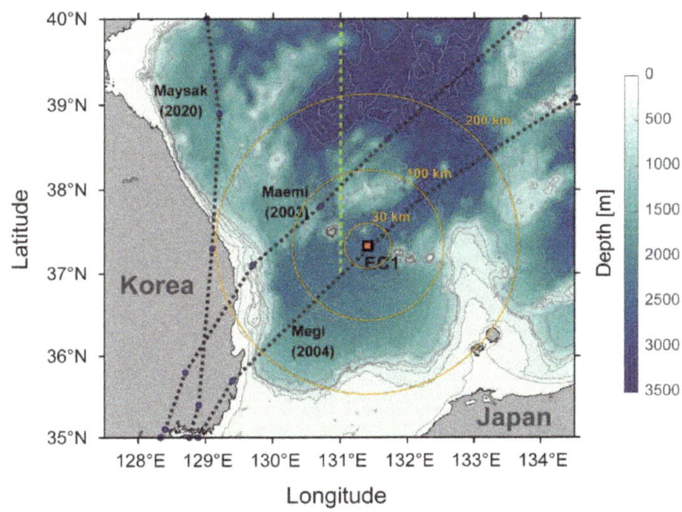

Figure 2. Location of subsurface mooring EC1 (red square) with bathymetry (colour) in the southwestern East Sea, off the east coast of Korea. Green dotted line indicates meridional line where sea surface wind data were extracted. Black dashed lines indicate trajectories of typhoon Maemi in 2003, Megi in 2004, and Maysak in 2020. The yellow circles indicate radii of 30, 100 and 200 km centred at EC1, respectively.

To supplement the moored time-series data at a horizontally fixed position, the Met Office Hadley Center EN4.2.1. (hereafter referred to as EN4) data that passed the global quality-control processing were used. The EN4 data consist of temperature and salinity obtained from profiling instruments, and the main data source is the World Ocean Database 2009 [42]. The temporal and spatial resolutions of these data are monthly and 1°, respectively. The spatial resolution of EN4 is coarser than that of the first baroclinic

Rossby radius of deformation of ~O (10 km), representing the horizontal scales at which the stratification can significantly vary within the EN4 grid [43]. Since the surface mixed layer depth (MLD) estimated from EN4 might not accurately represent the MLD for given spatial resolution, it was verified against the MLD estimated by Lim et al. [44], which were based on the World Ocean Database 2005 and multisource hydrographic data. Three vertical profiles of temperature observed using the profiling floats located within 80 km from EC1 were used to estimate the buoyancy frequency (Figure 3c,d). Satellite altimetry-derived daily sea-surface height (SSH; absolute dynamic topography above geoid) of gridded level four data provided by the Copernicus Marine Environment Monitoring Service were used to calculate the surface geostrophic currents, at a spatial resolution of 0.25°. The absolute dynamic topography was obtained by adding the mean dynamic topography to the sea level anomaly field and processed to provide the multimission merged altimeter data [45] used in this study. The mean dynamic topography is an estimate of the mean over 1993–2012 of the SSH above the geoid [45]. Hourly sea-surface wind data from January 2000 to November 2020 along a meridional line at 131° E (green dotted line in Figure 1), with a horizontal resolution of 30 km, were used to calculate the local surface wind stress, $\vec{\tau} = (\tau_x, \tau_y)$, retrieved from the European Centre for Medium-Range Weather Forecasts reanalysis version 5 (ECMWF, ERA5).

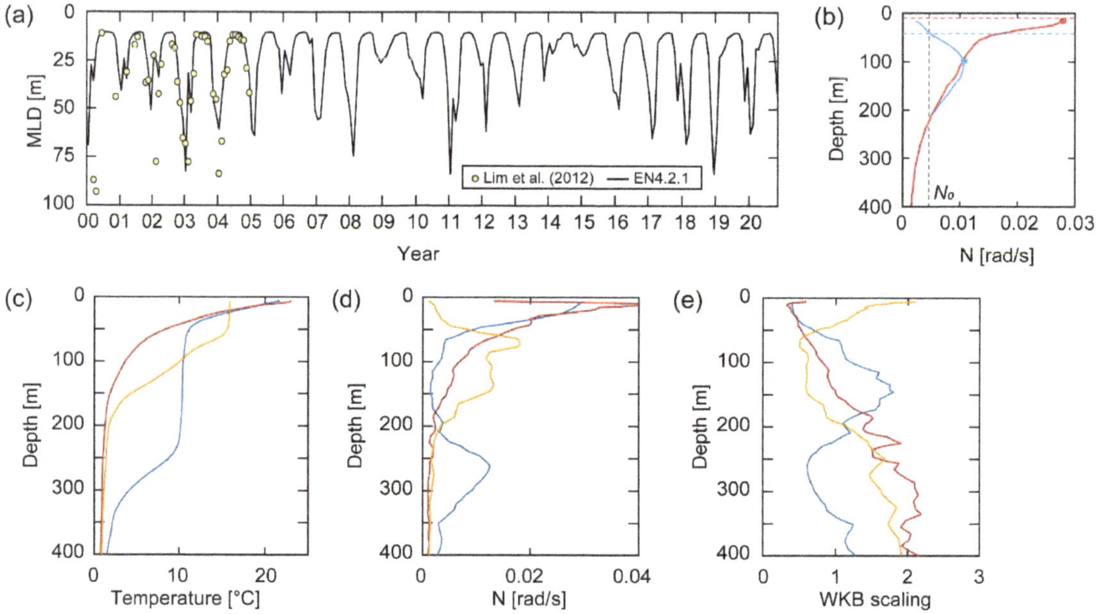

Figure 3. (**a**) Monthly mixed layer depth (MLD) estimated from EN4 (black line) data compared to that from Lim et al. [44] (yellow dot), (**b**) vertical structure of mean buoyancy frequencies, estimated from the EN4 data, at the upper 400 m nearby the EC1 mooring averaged over summer (June–July–August, red line) and winter (December–January–February, blue line) where the *Nmax* and MLD are remarked with circles and dashed horizontal lines with corresponding colours. Vertical dashed line in (**b**) indicates the depth-averaged reference N_0 (4.6×10^{-3} rad/s). Vertical profiles of (**c**) temperature, (**d**) buoyancy frequency and (**e**) WKB scaling factor observed using Argo floats located within 80 km from the EC1 (red, blue and orange lines were observed on 20 August 2014, 25 June 2015, and 23 December 2016, respectively.

2.2. Methods

Zonal and meridional currents (u, v) observed at 400 m of EC1 were processed to estimate near-inertial kinetic energy. Raw data extracted from the current meters were quality controlled by following the standard procedure for the instrument types [46,47] and then converted to hourly data through subsampling. The minimum current speed of 1.1 cm/s, measured using a rotor current meter (RCM), was treated as a stall and removed from successive processing. Consecutive missing data of less than 6 h were linearly interpolated, and those longer than 6 h were considered as bad data that were excluded from the analysis. Because the depths where the current meters were mounted differed from the nominal depths as the mooring was tilted by the drag due to horizontal currents, a linear interpolation (or extrapolation) was also performed vertically to yield the horizontal currents (u, v) at 400 m. The NIWs having zonal and meridional components of horizontal oscillations (u_{NIW}, v_{NIW}) were extracted from (u, v), by applying a phase-preserving fourth-order Butterworth bandpass filter, with cutoff frequencies of 0.85 f and 1.15 f; f was approximately 0.0505 cph, corresponding to a period of 19.8 h. The amplitude and horizontal kinetic energy of NIWs were computed as $\sqrt{u_{NIW}{}^2 + v_{NIW}{}^2}$ and $KE_{NIW\_obs} = 0.5\rho_0\left(u_{NIW}{}^2 + v_{NIW}{}^2\right)$, where ρ_0 is the reference density (=1025.0 kg/m^3). Because the density is not constant to the reference density but varies over time, the near-inertial potential energy as well as kinetic energy needs to be considered to represent the total mechanical energy.

Instead, to quantify the effect of NIW potential energy variations at 400 m, we deduced the time-varying Wentzel–Kramers–Brillouin (WKB) scaling factor as follows [48]:

$$[N(x,\ y,\ z,\ t)/N_0]^{-1/2}$$

where $N = [-(g/\rho_0)/(d\rho/dz)]^{1/2}$ is the buoyancy frequency, and x, y, z, t, N_0, and g are the zonal, meridional, and vertical coordinates, time, reference buoyancy frequency, and gravity acceleration (set to 9.83 m/s^2), respectively. Seasonal variations are clear and dominant in N at 50 m (mean and standard deviation are $11.9 \times 10^{-3} \pm 6.3 \times 10^{-3}$ rad/s) while decadal and longer-term changes are significant in N at 400 m (mean and standard deviation are $1.5 \times 10^{-3} \pm 2.4 \times 10^{-4}$ rad/s) (Figure 4a,c). In this study, N_0 was set to 4.6×10^{-3} rad/s, based on the EN4 data from the upper 500 m in the vicinity of EC1 from 2000 to 2020, by averaging the vertical profiles of buoyancy frequencies (e.g., vertical N profiles in summer and winter are compared to show the seasonal variation limited to the upper layers, Figure 3b). Three vertical N profiles obtained from the profiling floats within 80 km from EC1 were used to estimate the vertical profiles of WKB scaling factor (Figure 3c,d).

The intraseasonal variation in KE_{NIW_obs} was quantified by applying wavelet analysis to KE_{NIW_obs}, and intraseasonal band-averaged variance, defined as $\mathrm{VAR_{NIW\_obs\_int}}$, was extracted from the wavelet results. In our study, the intraseasonal band was set to 3–100 days, considering previously reported results on the mesoscale eddies in the region, for example, the mean lifetime of mesoscale eddies was estimated to be 95 days [49]. The MATLAB version of the wavelet toolbox [50] was used (http://atoc.colorado.edu/research/wavelets/, accessed on 12 November 2021). Events for high $\mathrm{VAR_{NIW\_obs\_int}}$ (period high) were defined as the criterion exceeding 1 standard deviation ($\sigma = 0.10$) from the mean ($\mu = 0.05$) of the $\mathrm{VAR_{NIW\_obs\_int}}$ normalized to its maximum (4.0×10^6 J^2/m^6) over the total period (i.e., $\mathrm{VAR_{NIW\_obs\_int}}$ exceeds $\sigma + \mu \sim 0.15$ for period high after the normalization; red dashed line in Figure 5c). Events for low $\mathrm{VAR_{NIW\_obs\_int}}$ (period low) were selected to match the total number of event days (505 days), the same as that of period high after sorting $\mathrm{VAR_{NIW\_obs\_int}}$ in order of magnitude, and the rest of the periods were defined as period neutral (Figure 5). A total of nine events of period high and seven events of period low were identified. Additionally, we defined the annual and decadal (2000s and 2010s) means of $\mathrm{VAR_{NIW\_obs\_int}}$ for the interannual and decadal variations in NIW kinetic energy.

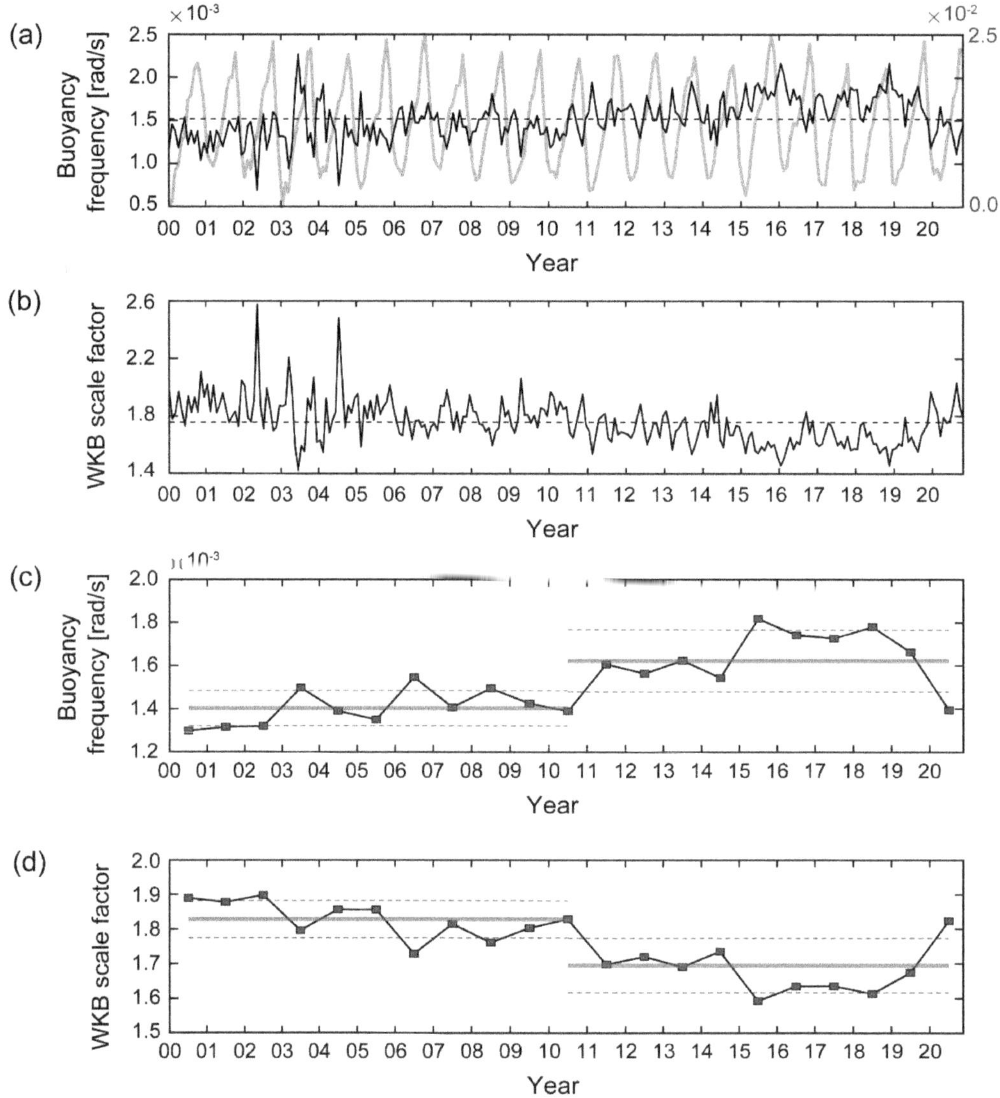

Figure 4. Time series of (**a**) monthly and (**c**) annual mean Buoyancy frequencies at 50 m (grey line, right y-axis in (**a**) only) and 400 m (black lines and squares, left y-axis) derived from EN4, (**b**) monthly and (**d**) annual mean Wentzel-Kramers-Brillouin (WKB) scale factors at 400 m. Dashed lines in (**a**,**b**) indicate temporal means for 400 m. Blue thick solid and thin dashed lines in (**c**,**d**) represent decadal means and their standard deviations for 2000s (from 2000 to 2010) and 2010s (from 2010 to 2020), respectively.

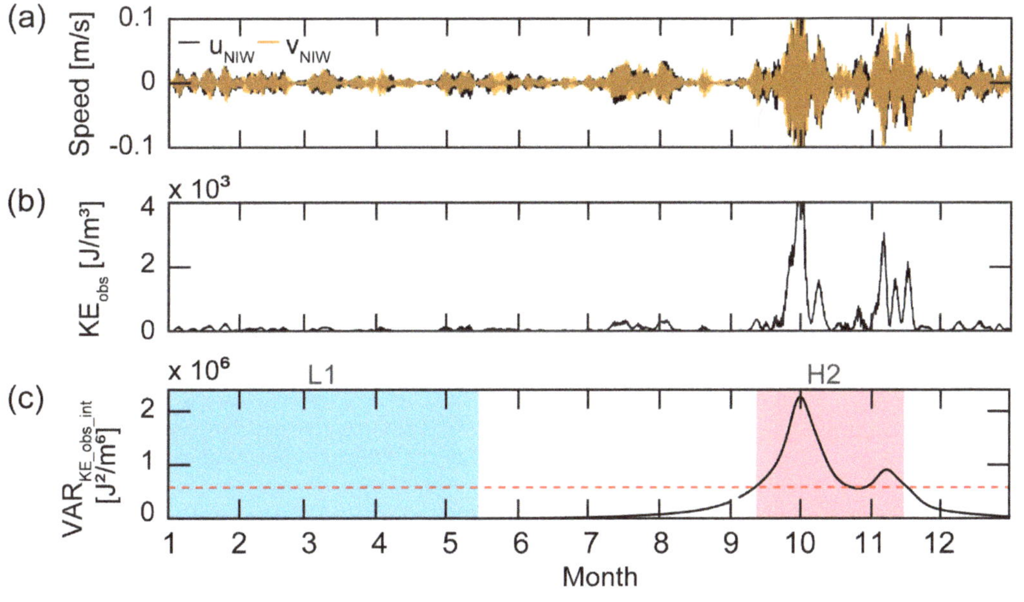

Figure 5. Time series of (**a**) zonal (u_{NIW}, black line) and meridional (v_{NIW}, orange line) components of horizontal NIW oscillations, (**b**) KE_{NIW_obs}, (c) $\mathrm{VAR}_{\mathrm{NIW\_obs\_int}}$ at 400 m in 2003 as an example. Red and blue shaded boxes and no shade in (**c**) indicate period high (H2), period low (L1), and period neutral, respectively. Red dashed line in (**c**) indicates $\sigma + \mu$.

The MLD is defined as the depth at which the density (ρ) changed by a given threshold criterion ($\Delta\rho$) relative to that at a reference depth [44,51]. The threshold was calculated from the temperature change (ΔT), relative to that at the reference depth, as follows:

$$\Delta\rho = \rho\left(T_{ref} + \Delta T,\ S_{ref},\ P_0\right) - \rho\left(T_{ref},\ S_{ref},\ P_0\right)$$

where T_{ref} and S_{ref} are the temperature and salinity at the reference depth derived from the EN4 data at the nearest grid to EC1, and P_0 is the pressure at the sea surface (set to zero). The threshold for temperature change and reference depth were set to $\Delta T = 0.2$ °C and 10 m, according to Lim et al. [44]. The σ and μ values of the MLD averaged over the period were 23 and 17 m, respectively (maximum of 84 m in January 2011 and minimum of 10.4 m in August 2006). The MLD varied seasonally, with the minimum and maximum values observed in summer (shallow MLD) and winter (deep MLD), respectively, and interannually with less (more) deepening in winters of 2008–2009 and 2014–2015 (2002–2003, 2010–2011, and 2018–2019) compared to other years (Figure 4). The MLDs estimated by Lim et al. [44] using the data collected within 1° distance from the EC1 mooring site were compared to validate the MLD estimated from the EN4 data; the correlation coefficient between the two MLD time series was 0.76.

To estimate the inertial response of the upper ocean in the mixed layer to surface wind forcing, a damped slab model (with zonal and meridional momentum equations) [52,53] was applied, using the following equations:

$$\frac{\partial u_{ML}}{\partial t} = f v_{ML} + \frac{\tau_x}{\rho_0 H_{ML}} - r u_{ML},\ \frac{\partial v_{ML}}{\partial t} = -f u_{ML} + \frac{\tau_y}{\rho_0 H_{ML}} - r v_{ML}$$

where H_{ML}, r^{-1}, (τ_x, τ_y), and (u_{ML}, v_{ML}) are the MLD, damping time scale, wind stress, and zonal and meridional currents in the mixed layer, respectively. The r^{-1} was fixed to 4 days, as considered in previous studies [36,37,54,55], and the time-varying MLD estimated from the EN4 data was used for determining H_{ML}.

The amplitude and kinetic energy of the modelled NIWs were calculated as $\sqrt{u_{ML}{}^2 + v_{ML}{}^2}$ and $KE_{NIW\_model} = 0.5\sigma_{\theta 0}(u_{ML}{}^2 + v_{ML}{}^2)$, respectively. The intraseasonal-band variance of KE_{NIW_model}, defined as $\mathrm{VAR}_{\mathrm{NIW\_model\_int}}$, was calculated in the same manner as that at 400 m by applying wavelet analysis. The rate of wind work (Π) was calculated using the following equation (inner product of surface wind stress and modelled mixed layer currents):

$$\Pi = \tau_{x_{NIW}} u_{ML\_NIW} + \tau_{y_{NIW}} u_{ML\_NIW}$$

where $\tau_{x_{NIW}}$ and $\tau_{y_{NIW}}$ are the near-inertial band-passed zonal and meridional wind stresses along the meridional line (see Figure 1). Note that the near-inertial band-passed currents (u_{ML_NIW}, v_{ML_NIW}) in the mixed layer estimated using the damped slab model (u_{ML}, v_{ML}) represent the near-inertial currents in the MLD to estimate Π.

The ray path of NIWs in the spatially varying stratification along 131° E (Figure 1) was computed as follows [56]:

$$\frac{2}{3}y^{3/2} = \frac{1}{(2\omega\beta)^{1/2}} \int N(y, z, t)dz$$

where y is the meridional travel distance, ω is the NIW frequency, N is the buoyancy frequency estimated using the EN4 data, and β is the meridional gradient of $f\,(= \partial f/\partial y \sim 1.8 \times 10^{-11}/\mathrm{m/s})$. The background flow fields were not considered in the calculation of the NIW ray path.

To examine the effect of the mesoscale flow fields on $\mathrm{VAR}_{\mathrm{NIW\_obs\_int}}$, background conditions were quantified from the satellite altimetry-derived surface geostrophic currents $\vec{U} = (U, V)$. The Okubo-Weiss parameter (α^2), which diagnoses the relative importance of the strain rate and relative vorticity, is defined as [57] follows:

$$\alpha^2 = \left(S_n{}^2 + S_s{}^2 - \zeta^2\right)/4$$

where S_n, S_s and ζ are the normal strain $\partial U/\partial x - \partial V/\partial y$, shear strain $\partial V/\partial x + \partial U/\partial y$, and relative vorticity $\partial V/\partial x - \partial U/\partial y$, respectively. When the total strain $S^2 = S_n{}^2 + S_s{}^2$ is larger than ζ^2, α^2 is positive, showing a saddle shape of the background flow fields. The effective Coriolis frequency was calculated as follows [27,58]:

$$f_{eff} = \sqrt{(f + \zeta/2)^2 - S^2/4}$$

To investigate whether there is a statistically significant effect of mesoscale flow fields on $\mathrm{VAR}_{\mathrm{NIW\_obs\_int}}$, composite analysis was performed by averaging (composite mean) values of Π, ζ, S^2, and α^2 separately for period high, period neutral, and period low, and compared to address whether they show statistically meaningful difference using Welch's t-test (with 95% significance level; p-value < 0.05) (Figure 6, Tables 1 and 2). Then, we classified the events into four categories based on the values of ζ and α^2 (Table 3). The dominance of S^2 to ζ^2 was determined by the sign of α^2 (e.g., when $\alpha^2 > 0$, the S^2 dominated ζ^2).

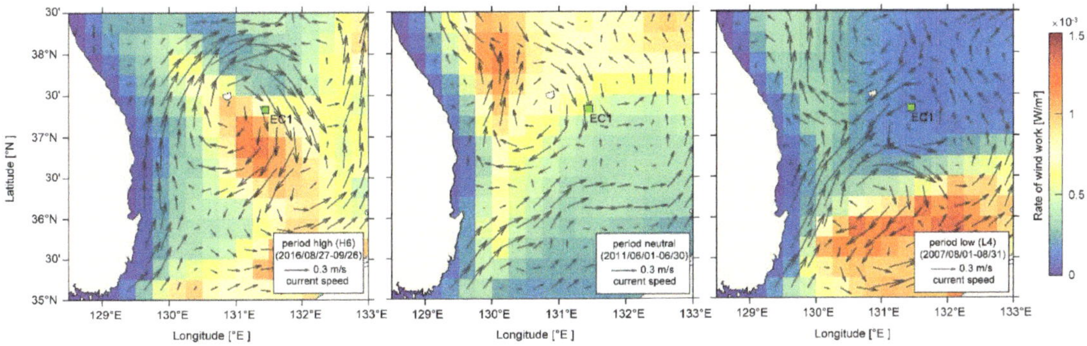

Figure 6. One-month-long composite maps of geostrophic current (grey arrow) superimposed on rate of wind work Π (in colours) during the period high (27 August–26 September 2016, H6), period neutral (1–30 June 2011), and period low (1–31 August 2007, L4). Green square marker represents the EC1.

Table 1. Mean and standard deviation (in bracket) of four condition parameters (Π, ζ, S^2, and α^2) during period high. Bold and underlined values are significant with 95% confidence (p-value < 0.05). For comparison, composite values for period neutral are shown in the bottom line.

	Period	Π (10^{-3} W/m^2)	ζ (f/s)	S^2 (f/s)	α^2 ($\times10^{-12}$/s^2)
H1	22 March–4 April 2000	0.39 (0.34)	−0.11 (0.02)	0.11 (0.03)	3.10 (9.89)
H2	12 September–15 November 2003	5.23 (23.52)	0.01 (0.04)	0.08 (0.03)	10.39 (10.92)
H3	12 March–5 April 2010	0.83 (1.17)	0.00 (0.02)	0.08 (0.02)	13.29 (6.29)
H4	12 June 2012–14 January 2013	0.71 (1.14)	−0.06 (0.03)	0.08 (0.02)	4.50 (12.45)
H5	23 February–18 June 2016	1.24 (4.11)	0.00 (0.08)	0.13 (0.03)	23.42 (26.68)
H6	27 August–26 September 2016	1.13 (2.82)	−0.10 (0.03)	0.06 (0.02)	−12.89 (17.01)
H7	20 October 2016–15 January 2017	1.16 (2.28)	−0.08 (0.06)	0.07 (0.03)	−5.90 (22.23)
H8	13–25 March 2018	0.07 (0.06)	0.04 (0.02)	0.04 (0.01)	0.58 (1.77)
H9	14 December 2019–26 April 2020	0.75 (1.35)	−0.07 (0.08)	0.06 (0.03)	−11.16 (19.09)
	Period High	1.54 (9.03)	−0.04 (0.08)	0.08 (0.04)	2.80 (24.01)
	Period Neutral	0.65 (2.27)	0.02 (0.10)	0.07 (0.04)	−4.58 (28.29)

Table 2. Same as Table 1 but during period low.

	Period	Π (10^{-3} W/m^2)	ζ (f/s)	S^2 (f/s)	α^2 ($\times10^{-12}$/s^2)
L1	30 November 2002–15 May 2003	0.32 (0.61)	0.09 (0.05)	0.04 (0.03)	−17.24 (23.45)
L2	29 January–24 February 2005	0.33 (0.60)	0.15 (0.01)	0.07 (0.01)	−33.99 (6.42)
L3	17–28 June 2005	0.27 (0.24)	−0.06 (0.01)	0.07 (0.01)	2.77 (3.47)
L4	14 August–16 September 2007	0.39 (1.08)	0.23 (0.02)	0.11 (0.03)	−80.73 (25.93)
L5	2 August–5 November 2008	0.48 (1.00)	0.03 (0.05)	0.06 (0.03)	2.41 (10.99)
L6	17 December 2008–14 April 2009	0.70 (1.15)	0.13 (0.10)	0.08 (0.03)	−33.41 (40.14)
L7	1 September–29 October 2009	0.22 (0.29)	0.07 (0.02)	0.04 (0.02)	−8.59 (7.57)
	Period Low	0.43 (0.87)	0.10 (0.08)	0.06 (0.03)	−20.45 (32.34)
	Period Neutral	0.65 (2.27)	0.02 (0.10)	0.07 (0.04)	−4.58 (28.29)

Table 3. Categories and corresponding events of $VAR_{NIW\_obs\_int}$ events using the two condition parameters of ζ and S^2 for mesoscale fields. The positive and negative anomalies are denoted by plus (+) and minus (−) signs referenced to a zero value for ζ and ζ^2 for S^2 during each event, respectively.

Category	ζ	S^2	Event
I	+	+	L5, H2, H8
II	+	−	L1, L2, L4, L6, L7, N1, N2
III	−	+	L3, H1, H3, H4, H5
IV	−	−	H6, H7, H9

3. Results

3.1. Intraseasonal, Interannual, and Decadal Variations of Near-Inertial Kinetic Energy below and within the Surface Mixed Layer

The $VAR_{NIW\_obs\_int}$ values showed significant intraseasonal, interannual, and decadal variations, rather than seasonal variations, from 2000 to 2020 (Figure 5 and Tables 1 and 2). Notably, $VAR_{NIW\_obs\_int}$ averaged over period high (red shaded box) was approximately 12 times higher than that of period neutral. During period low (blue shaded box), there was almost no $VAR_{NIW\_obs\_int}$ variation at 400 m (<0.02 J^2/m^6). The NIW kinetic energy (square root of $VAR_{NIW\_obs\_int}$) averaged over period high was ~1.1×10^3 J/m^3, which is approximately 24 times higher than that over period low (~4.7×10 J/m^3). Relatively high (>5.8×10^5 J^2/m^6) annual mean values of $VAR_{NIW\_obs\_int}$ were found in 2003, 2012–2013, 2016, and 2020, with the maximum value being observed in 2016 (with the peak value of 9.5×10^5 J^2/m^6 in 2016 corresponding to H5) (Figure 7b). In terms of decadal variations of $VAR_{NIW\_obs\_int}$, period high appeared more frequently in the 2010s than the 2000s, yielding a decadal mean of $VAR_{NIW\_obs\_int}$ in the 2010s (~8.3×10^4 J^2/m^6), significantly (95% confidence) higher than that in the 2000s (~3.1×10^5 J^2/m^6).

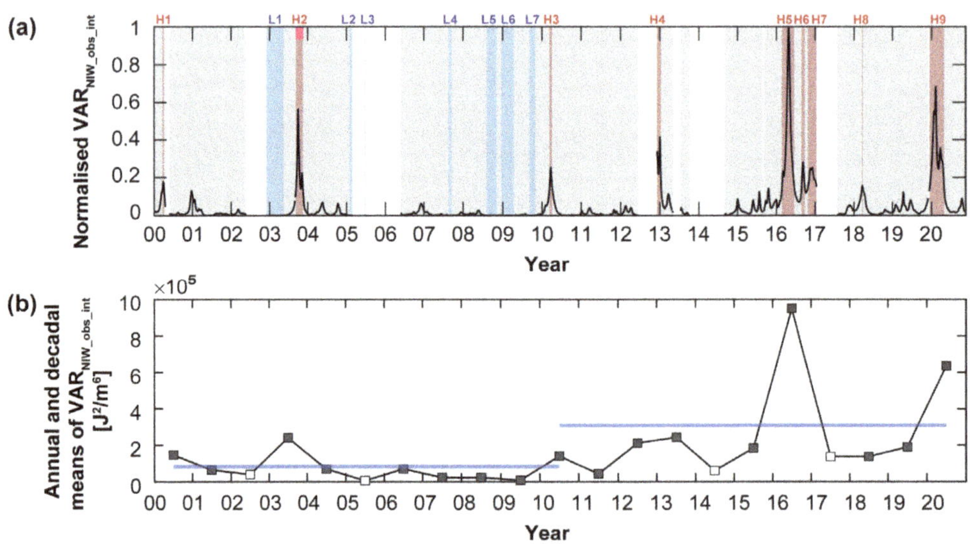

Figure 7. (**a**) Time series of normalised intraseasonal variations in KE_{NIW} at 400 m ($VAR_{NIW\_obs\_int}$) from 2000 to 2020. Red (blue) shaded boxes indicate period high (period low) and grey shaded area indicates period neutral. White area represents the data gap (no data available). Events 2–4 in Noh and Nam [40] are marked at the top of H2 (pink). (**b**) Annual (squares) and decadal mean (blue thick lines) of $VAR_{NIW\_obs\_int}$. In (**b**), grey and white squares indicate annual mean values of years in which the data acquisition rate was more and less than 50%.

Interestingly, the timing of the enhanced $VAR_{NIW\_model\_int}$ did not match well with that of $VAR_{NIW\_obs\_int}$ (Figures 7a and 8d,e). A $VAR_{NIW\_model\_int}$ (or amplitude of NIWs) value greater than 1.0×10^4 J^2/m^6 (~1 m/s) was found on 12 September 2003, 19 August 2004, and 3 September 2020 (green triangles in Figure 8d,e); notably, only the first date corresponded to period high (H2). H5 corresponded to the period when the value of $VAR_{NIW\_obs\_int}$ was higher than that during H2; however, the wind energy input during H5 (~7 kJ/m^2) was smaller than that during H2 (~31 kJ/m^2) (Figures 7a and 8c). Except for H2 and H5, the amplitudes of the modelled NIWs and $VAR_{NIW\_model\_int}$ were less than 0.2 m/s and 2.0×10^3 J^2/m, respectively. It is also interesting that there was a statistically significant difference in $VAR_{NIW\_model\_int}$ before and after 2010, yielding a higher decadal mean of 4.8×10^2 J^2/m^6 in the 2000s than the 4.5×10^2 J^2/m^6 observed in the 2010s, in contrast to the observational results (higher decadal mean of $VAR_{NIW\_obs\_int}$ in the 2010s; Figure 7b).

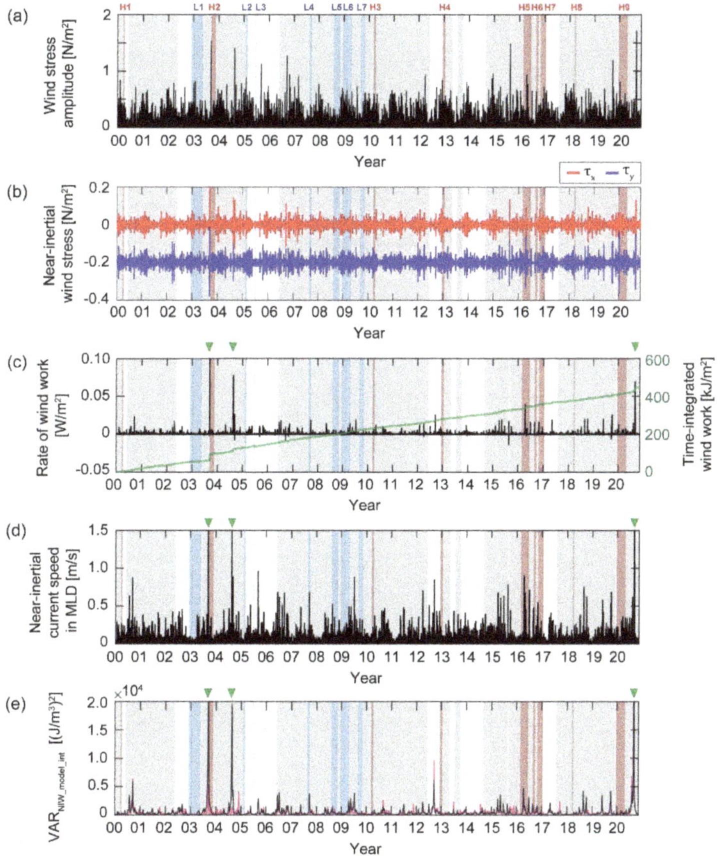

Figure 8. Time series of (**a**) wind stress amplitude, (**b**) near-inertial band-passed zonal (red) and meridional (blue) wind stresses, (**c**) rate of wind work Π (left axis) and time integral of Π showing cumulative wind energy input to the mixed layer (green solid line, right axis), (**d**) mixed layer near-inertial current amplitude, $\sqrt{u_{ML}{}^2 + v_{ML}{}^2}$, calculated from the damped slab model, and (**e**) intraseasonal-band variance of (**d**) from local (black) and remote (pink) wind stresses where the latter denotes wind stress averaged over 38–40° N. Red (blue) shaded boxes indicate period high (period low). Grey shaded area indicates period neutral. Green triangles in (**c**–**e**) indicate the periods when the mixed layer near-inertial current amplitude was larger than 1 m/s.

3.2. Composite Mean of Near-Inertial Kinetic Energy at 400 m Dependent on Mesoscale Condition

In our study, the magnitudes of Π and S^2 composited for period high (period low) were significantly larger (smaller) than those for period neutral, showing 1.54×10^{-3} W/m^2 (0.43×10^{-3} W/m^2) and $0.08f$/s ($0.06f$/s), respectively (Figures 8 and 9 and Tables 1 and 2). The Π during the period high events (except for H1 and H8) were significantly larger than the composite mean during period neutral, while those during period low events (except L6) were significantly smaller than those during period neutral. The α^2 and ζ composite mean values for period high (period low) had positive (negative) and negative (positive) signs, showing $+2.80 \times 10^{-12}/\text{s}^2$ ($-20.45 \times 10^{-12}/\text{s}^2$) and $-0.04f$/s ($0.10f$/s), indicating the dominance of strain to vorticity (vorticity to strain) and lower (higher) f_{eff}, respectively. At period high, it was shown that the $\zeta < 0$ or strain fields were strengthened by mesoscale flow fields, while at period low, $\zeta > 0$ by cyclonic circulation appeared (Figure 6). More than half of the period high and period low events could be identified by the signs of ζ and α^2. The role of the ζ was identified by comparing Categories I and III for $S^2 > \zeta^2$ (corresponding to $\alpha^2 > 0$), and Categories II and VI for $S^2 < \zeta^2$ (corresponding to $\alpha^2 < 0$), commonly yielding more period high events for a negative ζ (Table 3). The role of S^2 was determined by comparisons between Categories I and II for $\zeta > 0$ and between Categories III and IV for $\zeta < 0$, commonly yielding more period high events for $S^2 > \zeta^2$ (positive α^2). For each Category, the composite means of $\text{VAR}_{\text{NIW\_obs\_int}}$ were 1.05×10^5, 1.29×10^5, 1.25×10^5, and 1.43×10^5 J^2/m^6 yielding that at Category I < Category II, Category III < Category < IV and Category II < Category IV with 95% significant level (p-value < 0.05). There was no significantly high correlation between Category II and Category III (p-value > 0.05).

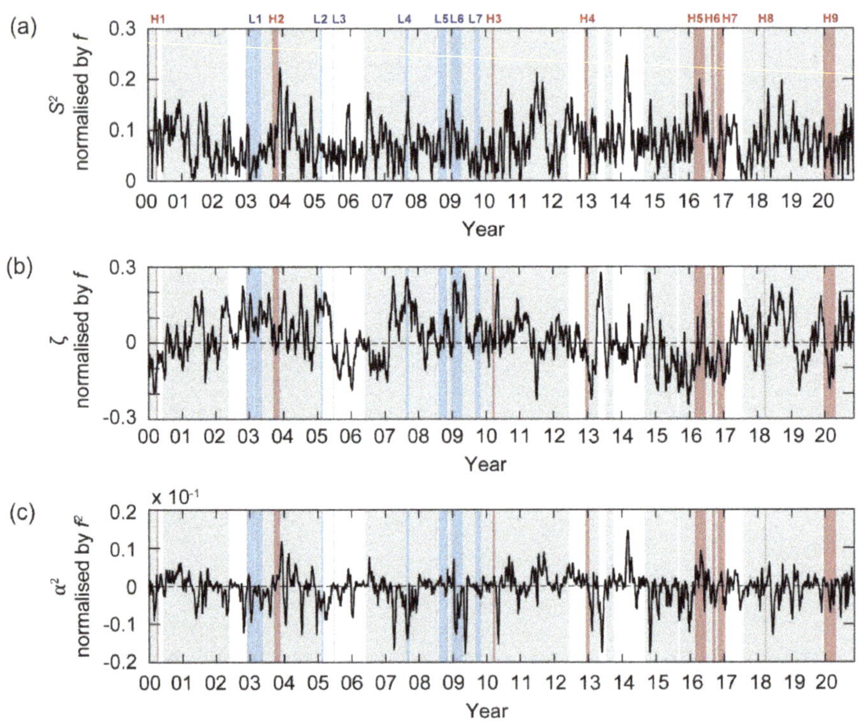

Figure 9. Time series of (**a**) S^2 normalised by f, (**b**) ζ normalised by f, and (**c**) α^2 normalised by f^2. Red (blue) shaded boxes indicate period high (period low). Grey shaded area indicates period neutral.

4. Discussion

4.1. Comparison to Known Characteristics of Near-Inertial Internal Waves (NIWs)

We compared the results of nonseasonal variations of the observed and modelled NIW kinetic energy ($VAR_{NIW\_obs\_int}$) with previously known characteristics of NIWs. In general, period high often corresponded to the condition of $\zeta < 0$ (Category III or IV), supporting that NIWs are trapped in a region of $\zeta < 0$ because of the lowered f_{eff} due to mesoscale circulations, as suggested in previous studies [37,38]. The interannual variations in the region of $\zeta < 0$ imposed by the meandering of the subpolar front and the activities of anticyclonic eddies off the east coast of Korea may be responsible for the interannual variations of NIW kinetic energy presented in [38] and the horizontal and vertical distributions of NIW kinetic energy shown in [37]. The NIWs presented in [39,59] under the condition of $\zeta > 0$ corresponded to Categories I or II in our study. Notably, significant impacts of S^2 on the NIW kinetic energy in the region have not been reported previously, except for a study conducted by Noh and Nam [40], which suggested a strong case of NIWs extracting their energy from a mesoscale flow field under the favourable condition of $S^2 > \zeta^2$ (strain exceeds vorticity, $\alpha^2 > 0$) via the wave capture process [32,33]. The case found during events 2–4 in the study conducted by Noh and Nam [40], which supported the exponentially growing wavenumber, along with an increasing S^2 and a resulting small group velocity, corresponded to H2 in our study; this was generalised from statistically significant differences in the composite mean variance of NIW kinetic energy, for example, more period high events for $S^2 > \zeta^2$.

Wind-induced NIWs are known to propagate downward below the MLD [37,60,61]. However, in spite of large Π values, only $VAR_{NIW\_model\_int}$ (and not $VAR_{NIW\_obs\_int}$) showed an occasional enhancement, such as that observed on 19 August 2004 and 3 September 2020 (green triangles in Figure 8d), when the typhoon passed the region. This means that the nonseasonal variations in $VAR_{NIW\_obs\_int}$ could not be well-explained by the changes in the surface wind forcing alone, additionally requiring the consideration of mesoscale field conditions. Another circumstantial energy source for NIWs in the deep layer in the region suggested by Mori et al. [36] is topographic roughness, because barotropic currents flowing over the rough bottom could generate deep near-inertial oscillations. Indeed, in this case, the current-topography interaction will also be affected by the mesoscale circulation, because the mesoscale eddies in the region are quasi-barotropic, as discussed in [39], which accounts for the higher NIW kinetic energy observed inside (where $\zeta < 0$) the anticyclonic circulation in the region.

4.2. Effects of Surface Wind Forcing on Near-Inertial Energy within and below the Mixed Layer

The NIWs within the mixed layer can be easily amplified by the changes in the local wind stress; however, the wind-induced NIWs within or just below the mixed layer may dissipate mostly (up to 70–85%) in the upper 200 m [61–63], and hardly penetrate far below the MLD to 400 m depth, accounting for significantly different nonseasonal variations in the KE_{NIW_obs} at 400 m of the EC1 ($VAR_{NIW\_obs\_int}$) compared to those in KE_{NIW_model}, within the mixed layer simulated by the damped slab model using local wind stress ($VAR_{NIW\_model\_int}$). Positive anomalies of $VAR_{NIW\_model\_int}$ exceeding $\sigma + \mu$ ($\sim 2.5 \times 10^3$ J^2/m^6) were not found during the events of period high, although they were clearly accompanied by relatively high Π values during H2 and H5 (Figures 7 and 8). Indeed, periods of very high $VAR_{NIW\_model\_int}$ (19 August 2004 and 3 September 2020) exceeding 1.0×10^4 J^2/m^6 in association with the passage of typhoons (Figure 2) corresponded to period neutral events indicative of only moderate $VAR_{NIW\_obs\_int}$. In general, $VAR_{NIW\_model\_int}$ was low (less than 2.0×10^3 J^2/m^6) when Π decreased during period low. Significant dissipation of wind-induced NIWs may be confirmed around August 2004 (period neutral, green triangle in Figure 8); for example, the case when the KE_{NIW_obs} of 1.12×10^2 J/m^3 observed at 400 m (corresponding to the square root of peak $VAR_{NIW\_obs\_int}$ of 1.26×10^4 J^2/m^6) explained only 9.6% of the KE_{NIW_model} of ~11.68×10^2 J/m^3 within the mixed layer, supporting the deduction or strong dissipation of NIWs in the upper 200 m [62,63].

Because the surface wind stress is not uniform, equatorward- and downward-propagating NIWs generated by strong wind forcing in the north of EC1, despite the weak local wind forcing, may propagate down to 400 m of EC1, accounting for the high $VAR_{NIW\_obs\_int}$ without a significant $VAR_{NIW\_model\_int}$ (except H2 and H5). However, this possibility can be ruled out, as the intraseasonal variance in the kinetic energy of NIWs originating from higher latitudes (forced by wind stress at 38–40° N along the 131° E), propagating below the mixed layer to 400 m (of EC1; Figure 10), was not markedly different from $VAR_{NIW\_model\_int}$, nor did it correlate with $VAR_{NIW\_obs\_int}$ (Figure 8e).

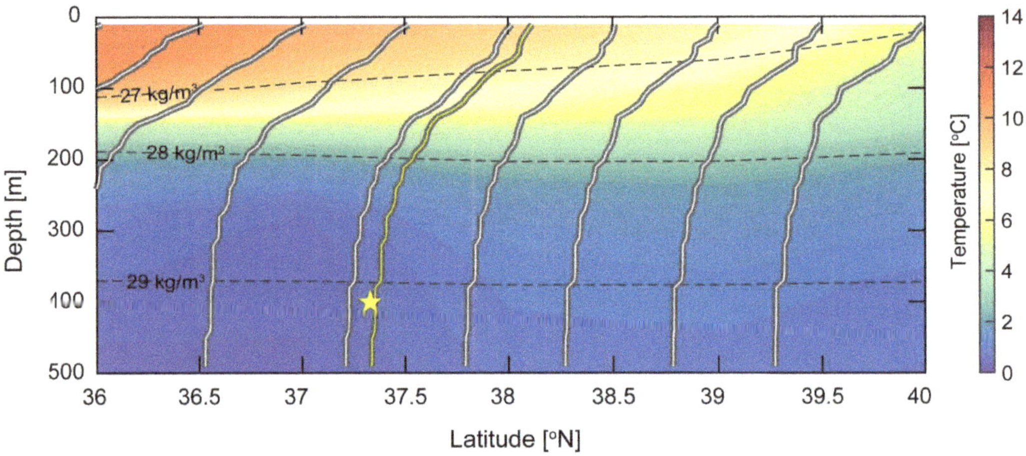

Figure 10. Ray path of near-inertial internal waves (NIWs) (thick, white, solid line) in the spatially varying stratification observed in March 2018 (H8). Yellow solid line indicates ray path that could reach the depth of 400 m at EC1 (yellow star). Background colour indicates potential temperature, and dashed line indicates isopycnals of 27, 28, and 29 kg/m^3.

4.3. *Effects of Mesoscale Flow Fields on Near-Inertial Energy Far below the Mixed Layer*

The condition parameters of ζ and S^2 that represent the mesoscale flow field explained $VAR_{NIW\_obs\_int}$ better than surface wind forcing, because a negative ζ (Categories III and IV) lowers f_{eff} (permitting the trapping and rapid deep propagation of NIWs) (Figure 6) [24,28] and increases S^2, leading to a positive anomaly of α^2 (Categories I and III), which stretches and rotates the wavevector (resulting in exponentially growing wavenumber and decreasing group velocity), representing an enhancement of NIWs [30]. For example, the lowered f_{eff} ($\zeta < 0$) associated with the anticyclonic circulation around the EC1 supports the high $VAR_{NIW\_obs\_int}$ during the H6, contrasting to the low $VAR_{NIW\_obs\_int}$ during the L4 when $\zeta > 0$, associated with the cyclonic circulation around the EC1 (Figure 6). During the two events (N1 and N2) of the neutral period, we observed low $VAR_{NIW\_obs\_int}$ values despite high $VAR_{NIW\_model\_int}$ values, in association with the typhoon passage, which may be explained by unfavourable conditions ($\zeta > 0$ and $S^2 < \zeta^2$; Category II) for NIW enhancement imposed by mesoscale conditions (Table 3). In contrast, more period high events were observed under favourable conditions ($\zeta < 0$ and $S^2 > \zeta^2$; Category III) for NIW enhancement imposed by mesoscale conditions (Table 3), mainly regardless of wind forcing.

The results of more frequent events of period high in the 2010s (compared to those in the 2000s) were also accounted for by the mesoscale field condition, and not by the surface-wind-induced NIWs within the mixed layer. The anticyclonic UWE anomalously lasted longer (nearly two years from October 2014 to August 2016) in the region [64], and a newly formed UWE appeared again in September 2016, providing mesoscale conditions of $\zeta < 0$ that were favourable for NIW enhancement at 400 m of EC1 during H5, H6, and H7. Changes in mesoscale flow conditions due to the UWE in the 2010s accompanied the strengthening of density stratification in the region, yielding significantly higher N and lower WKB scaling factor values at 400 m of EC1 during the 2010s, compared to those observed in the 2000s, due to a strong stratification linked to mesoscale conditions in 2003, 2006, 2011, 2013, and 2015–2018 (Figure 4). Interestingly, the 14% decrease in the WKB scaling factor or the increase in the NIW potential energy in the 2010s (compared to that observed in the 2000s), however, could not explain the increased frequency of period high events in the 2010s. Note that the events of period high and period low were not significantly dependent on whether WKB scaling was applied (now shown), indicating that the nonseasonal variation of NIW energy was more affected by the mesoscale flow field (vorticity and strain) than the stratification. The WKB scaling factor at 400 m was variable depending on the varying stratification, but always higher than the unity at the depth (Figures 3c–e and 4b,d).

5. Concluding Remarks

Nonseasonal (intraseasonal, interannual, and decadal) variations in NIW kinetic energy, KE_{NIW_obs} at 400 m ($VAR_{NIW\_obs\_int}$) were presented from long time-series (from 2000 to 2020) moored observations in the southwestern East Sea. In total, nine periods of high (period high) and seven periods of low (period low) intraseasonal variance of KE_{NIW_obs}, or $VAR_{NIW\_obs\_int}$ at 400 m, were identified and analysed statistically, providing composite means for different mesoscale conditions, and suggesting a significant effect of mesoscale flow fields on $VAR_{NIW\_obs\_int}$. Although a high rate of wind work, sometimes associated with typhoon passage, may significantly enhance the near-inertial kinetic energy (KE_{NIW_model} within or just below the surface mixed layer as simulated by the damped slab model), in our study, the intraseasonal variance of KE_{NIW_model}, or $VAR_{NIW\_model\_int}$, hardly accounted for $VAR_{NIW\_obs\_int}$ at 400 m depth, which was dissipated mostly in the upper layer. Instead, the condition parameters of relative vorticity (ζ) and the total strain (S^2), representing mesoscale flow fields, better explained $VAR_{NIW\_obs\_int}$ at 400 m, yielding a significantly higher KE_{NIW_obs} at 400 m when $\zeta < 0$ and/or $S^2 > \zeta^2$, and vice versa (Figure 11). Our results, based on 21-year-long observations, statistically support previous (mostly theoretical) suggestions that NIW kinetic energy may have been enhanced through nonlinear interactions with the mesoscale flow field, when the strain exceeded the vorticity, and was trapped when the effective Coriolis frequency was lowered due to a negative relative vorticity. We believe that future process-oriented studies focusing on specific NIW events will provide a more comprehensive understanding of the interaction between the mesoscale flow fields and NIWs, testing the statistically derived proposition obtained in this study from rare, long-term, high-resolution, and continuous observations.

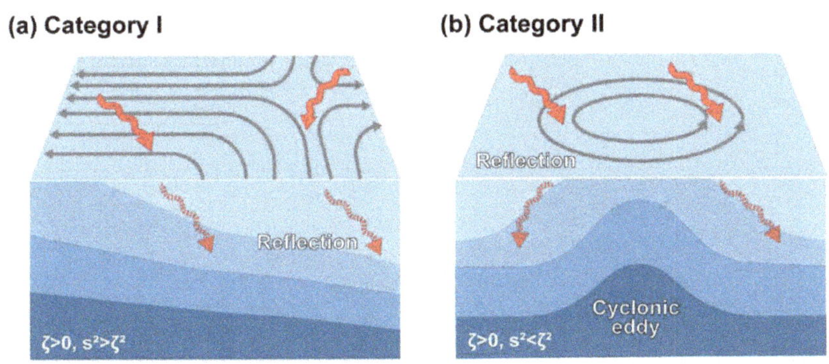

Figure 11. *Cont.*

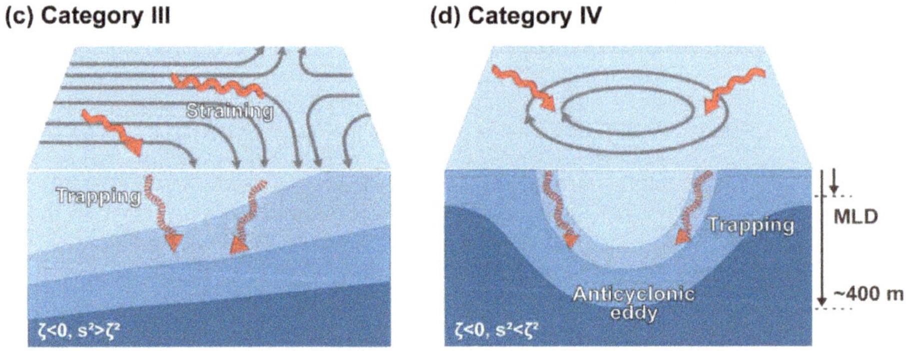

Figure 11. Schematics of behaviour of near-inertial internal waves (NIW, red arrow) under mesoscale flow fields (streamlines, grey arrows) yielding different conditions of ζ and α^2; (**a**) Category I ($\zeta > 0$ and $S^2 > \zeta^2$), (**b**) Category II ($\zeta > 0$ and $S^2 < \zeta^2$), (**c**) Category III ($\zeta < 0$ and $S^2 > \zeta^2$), and (**d**) Category IV ($\zeta < 0$ and $S^2 < \zeta^2$). Different blue shading colours represent different isopycnal layers. Vertical scales of MLD and 400 m reference level are shown with thick grey arrows in the right bottom of (**d**).

Author Contributions: Conceptualisation, S.N. (Suyun Noh) and S.N. (SungHyun Nam); methodology, S.N. (Suyun Noh) and S.N. (SungHyun Nam); validation, S.N. (Suyun Noh) and S.N. (SungHyun Nam); formal analysis, S.N. (Suyun Noh); investigation, S.N. (Suyun Noh) and S.N. (SungHyun Nam); data curation, S.N. (Suyun Noh); writing—original draft preparation, S.N. (Suyun Noh); writing—review and editing, S.N. (Suyun Noh) and S.N. (SungHyun Nam); visualisation, S.N. (Suyun Noh); supervision, S.N. (SungHyun Nam); project administration, S.N. (SungHyun Nam). All authors have read and agreed to the published version of the manuscript.

Funding: This research was funded by the Ministry of Oceans and Fisheries, Republic of Korea through the Deep Water Circulation and Material Cycling in the East Sea (20160040). This work was also partially funded by the Civil Military Technology Cooperation Program (18-SN-RB-01) from the Institute of Civil Military Technology Cooperation, Republic of Korea.

Institutional Review Board Statement: Not applicable.

Informed Consent Statement: Not applicable.

Data Availability Statement: All EC1 data are available at SEANOE (https://www.seanoe.org/data/00677/78916/, accessed on 12 November 2021) and OceanSITES (http://www.oceansites.org, accessed on 12 November 2021). Satellite altimetry-derived daily sea-surface height of gridded level 4 data provided by the Copernicus Marine Environment Monitoring Service are available at https://marine.copernicus.eu/ (accessed on 12 November 2021). The sea surface wind data, ERA5, were provided by the European Center for Medium-Range Weather Forecasts (https://www.ecmwf.int/en/forecasts/datasets/reanalysis-datasets/era5, accessed on 12 November 2021). The EN4 dataset was collected and provided by the Met Office Hadley Centre (https://www.metoffice.gov.uk/hadobs/en4/, accessed on 12 November 2021).

Acknowledgments: We would like to thank two reviewers for constructive comments, and Jae-Hun Park and Sehan Lim who generously provided the hydrographic data required for our study, along with their thoughtful insights on the preliminary results.

Conflicts of Interest: The authors declare no conflict of interest.

References

1. Garrett, C. What is the "near-inertial" band and why is it different from the rest of the internal wave spectrum? *J. Phys. Oceanogr.* **2001**, *31*, 962–971. [CrossRef]
2. Alford, M.H.; MacKinnon, J.A.; Simmons, H.L.; Nash, J.D. Near-inertial internal gravity waves in the ocean. *Annu. Rev. Mar. Sci.* **2016**, *8*, 95–123. [CrossRef]

3. MacKinnon, J.A.; Alford, M.H.; Ansong, J.K.; Arbic, B.K.; Barna, A.; Briegleb, B.P.; Bryan, F.O.; Buijsman, M.C.; Chassignet, E.P.; Danabasoglu, G.; et al. Climate process team on internal wave-driven ocean mixing. *Bull. Am. Meteorol. Soc.* **2017**, *98*, 2429–2454. [CrossRef] [PubMed]
4. Whalen, C.B.; MacKinnon, J.A.; Talley, L.D. Large-scale impacts of the mesoscale environment on mixing from wind-driven internal waves. *Nat. Geosci.* **2018**, *11*, 842–847. [CrossRef]
5. Egbert, G.D.; Ray, R.D. Significant dissipation of tidal energy in the deep ocean inferred from satellite altimeter data. *Nature* **2000**, *405*, 775–778. [CrossRef]
6. Nycander, J. Generation of internal waves in the deep ocean by tides. *J. Geophys. Res.* **2005**, *110*, C10028. [CrossRef]
7. Jiang, J.; Lu, Y.; Perrie, W. Estimating the energy flux from the wind to ocean inertial motions: The sensitivity to surface wind fields. *Geophys. Res. Lett.* **2005**, *32*, L15610. [CrossRef]
8. Garrett, C.J.R.; Munk, W. Internal waves in the ocean: A progress report. *Annu. Rev. Fluid Mech.* **1979**, *11*, 339–369. [CrossRef]
9. Munk, W.; Wunsch, C. Abyssal recipes II: Energetics of tidal and wind mixing. *Deep-Sea Res. Part I Oceanogr. Res. Pap.* **1998**, *45*, 1977–2010. [CrossRef]
10. Müller, P.; Briscoe, M. Diapycnal mixing and internal waves. *Oceanography* **2000**, *13*, 98–103. [CrossRef]
11. Alford, M.H. Redistribution of energy available for ocean mixing. *Nature* **2003**, *428*, 159–162. [CrossRef]
12. Ferrari, R.; Wunsch, C. Ocean circulation kinetic energy: Reservoirs, sources, and sinks. *Annu. Rev. Fluid Mech.* **2009**, *41*, 253–282. [CrossRef]
13. Whalen, C.B.; de Lavergne, C.; Naveira Garabato, A.C.; Klymak, J.M.; MacKinnon, J.A.; Sheen, K.L. Internal wave-driven mixing: Governing processes and consequences for climate. *Nat. Rev.* **2020**, *1*, 606–621. [CrossRef]
14. Nikurashin, M.; Ferrari, R. Overturning circulation driven by breaking internal waves in the deep ocean. *Geophys. Res. Lett.* **2013**, *40*, 3133–3137. [CrossRef]
15. Vic, C.; Garabato, A.C.N.; Green, J.M.; Waterhouse, A.F.; Zhao, Z.; Melet, A.; De Lavergne, C.; Buijsman, M.C.; Stephenson, G.R. Deep-ocean mixing driven by small-scale internal tides. *Nat. Commun.* **2019**, *10*, 2099. [CrossRef]
16. Blaker, A.T.; Hirschi, J.J.M.; Sinha, B.; de Cuevas, B.; Alderson, S.; Coward, A.; Madec, G. Large near-inertial oscillations of the Atlantic meridional overturning circulation. *Ocean Model.* **2012**, *42*, 50–56. [CrossRef]
17. Xiao, J.; Xie, Q.; Wang, D.; Yang, L.; Shu, Y.; Liu, C.; Chen, J.; Yao, J.; Chen, G. On the near-inertial variations of meridional overturning circulation in the South China Sea. *Ocean Sci.* **2016**, *12*, 335–344. [CrossRef]
18. Lucas, A.J.; Franks, P.J.S.; Dupont, C.L. Horizontal internal-tide fluxes support elevated phytoplankton productivity over the inner continental shelf. *Limnol. Oceanog.* **2011**, *1*, 56–74. [CrossRef]
19. Pan, X.; Wong, G.T.F.; Shiah, F.-K.; Ho, T.-Y. Enhancement of biological productivity by internal waves: Observations in the summertime in the northern South China Sea. *J. Oceanogr.* **2012**, *68*, 427–437. [CrossRef]
20. Muacho, S.; da Silva, J.C.B.; Brotas, V.; Oliveira, P.B. Effect of internal waves on near-surface chlorophyll concentration and primary production in the Nazaré Canyon (west of the Iberian Peninsula). *Deep Sea Res. Part I Oceanogr. Res. Pap.* **2013**, *81*, 89–96. [CrossRef]
21. Villamaña, M.; Mouriño-Carballido, B.; Marañón, E.; Cermeño, P.; Chouciño, P.; da Silva, J.C.B.; Díaz, P.A.; Fernández-Castro, B.; Gilcoto, M.; Graña, R.; et al. Role of internal waves on mixing, nutrient supply and phytoplankton community structure during spring and neap tides in the upwelling ecosystem of Ría de Vigo (NW Iberian Peninsula). *Limnol. Oceanogr.* **2017**, *62*, 1014–1030. [CrossRef]
22. Li, D.; Chou, W.C.; Shih, Y.Y.; Chen, G.Y.; Chang, Y.; Chow, C.H.; Lin, T.Y.; Hung, C.C. Elevated particulate organic carbon export flux induced by internal waves in the oligotrophic northern South China Sea. *Sci. Rep.* **2018**, *8*, 2042. [CrossRef] [PubMed]
23. Song, H.; Marshall, J.; Campin, J.M.; McGillicuddy, D.J. Impact of near-inertial waves on vertical mixing and air-sea CO_2 fluxes in the Southern Ocean. *J. Geophys. Res. Oceans* **2019**, *124*, 4605–4617. [CrossRef]
24. Kunze, E. Near-Inertial Wave Propagation in geostrophic shear. *J. Phys. Oceanogr.* **1985**, *15*, 544–565. [CrossRef]
25. D'asaro, E.A. Upper-ocean inertial currents forced by a strong storm. Part III: Interaction of inertial currents and mesoscale eddies. *J. Phys. Oceanogr.* **1995**, *25*, 2953–2958. [CrossRef]
26. Whitt, D.B.; Thomas, L.N. Resonant generation and energetics of wind-forced near-inertial motions in a geostrophic flow. *J. Phys. Oceanogr.* **2015**, *45*, 181–208. [CrossRef]
27. Jing, Z.; Wu, L.; Ma, X. Energy exchange between the mesoscale oceanic eddies and wind-forced near-inertial oscillations. *J. Phys. Oceanogr.* **2017**, *47*, 721–733. [CrossRef]
28. Lee, D.-K.; Niiler, P.P. The inertial chimney: The near-inertial energy drainage from the ocean surface to the deep layer. *J. Geophys. Res. Oceans* **1998**, *103*, 7579–7591. [CrossRef]
29. Danioux, E.; Klein, P.; Rivière, P. Propagation of wind energy into the deep ocean through a fully turbulent mesoscale eddy field. *J. Phys. Oceanogr.* **2008**, *38*, 2224–2241. [CrossRef]
30. BÜhler, O.; McIntyre, M.E. Wave capture and wave–vortex duality. *J. Fluid Mech.* **2005**, *534*, 67–95. [CrossRef]
31. Polzin, K.L. Mesoscale eddy–internal wave coupling. Part I: Symmetry, wave Capture, and results from the mid-ocean dynamics experiment. *J. Phys. Oceanogr.* **2008**, *38*, 2556–2574. [CrossRef]
32. Polzin, K.L. Mesoscale eddy–internal wave coupling. Part II: Energetics and results from PolyMode. *J. Phys. Oceanogr.* **2010**, *40*, 789–801. [CrossRef]

33. Jing, Z.; Chang, P.; DiMarco, S.F.; Wu, L. Observed energy exchange between low-frequency flows and internal waves in the Gulf of Mexico. *J. Phys. Oceanogr.* **2018**, *48*, 995–1008. [CrossRef]
34. Chang, K.I.; Teague, W.J.; Lyu, S.J.; Perkins, H.T.; Lee, D.K.; Watts, D.R.; Kim, Y.B.; Mitchell, D.A.; Lee, C.M.; Kim, K. Circulation and currents in the southwestern East/Japan Sea: Overview and review. *Prog. Oceanogr.* **2004**, *61*, 105–156. [CrossRef]
35. Shin, H.-R.; Shin, C.-W.; Kim, C.; Byun, S.-K.; Hwang, S.-C. Movement and structural variation of warm eddy WE92 for three years in the Western East/Japan Sea. *Deep Sea Res. Part II Top. Stud. Oceanogr.* **2005**, *52*, 1742–1762. [CrossRef]
36. Mori, K.; Matsuno, T.; Senjyu, T. Seasonal/spatial variations of the near-inertial oscillations in the deep water of the Japan Sea. *J. Oceanogr.* **2005**, *61*, 761–773. [CrossRef]
37. Jeon, C.; Park, J.H.; Park, Y.G. Temporal and spatial variability of near-inertial waves in the East/Japan Sea from a high-resolution wind-forced ocean model. *J. Geophys. Res. Oceans.* **2019**, *124*, 6015–6029. [CrossRef]
38. Park, J.-H.; Watts, D.R. Near-inertial oscillations interacting with mesoscale circulation in the southwestern Japan/East Sea. *Geophys. Res. Lett.* **2005**, *32*, L10611. [CrossRef]
39. Byun, S.-S.; Park, J.J.; Chang, K.-I.; Schmitt, R.W. Observation of near-inertial wave reflections within the thermostad layer of an anticyclonic mesoscale eddy. *Geophys. Res. Lett.* **2010**, *37*, L01606. [CrossRef]
40. Noh, S.; Nam, S. Observations of enhanced internal waves in an area of strong mesoscale variability in the southwestern East Sea (Japan Sea). *Sci. Rep.* **2020**, *10*, 9068. [CrossRef] [PubMed]
41. Lee, H.; Nam, S. *EC1, Long and Continuous Mooring Time Series from 1996 to 2020*; SEANOE: Paris, France, 2021. [CrossRef]
42. Good, S.A.; Martin, M.J.; Rayner, N.A. EN4: Quality controlled ocean temperature and salinity profiles and monthly objective analyses with uncertainty estimates. *J. Geophys. Res. Oceans* **2013**, *118*, 6704–6716. [CrossRef]
43. Chelton, D.B.; DeSzoeke, R.A.; Schlax, M.G.; El Naggar, K.; Siwertz, N. Geographical variability of the first baroclinic Rossby radius of deformation. *J. Phys. Oceanogr.* **1998**, *28*, 433–460. [CrossRef]
44. Lim, S.; Jang, C.J.; Oh, I.S.; Park, J. Climatology of the mixed layer depth in the East/Japan Sea. *J. Mar. Syst.* **2012**, *96–97*, 1–14. [CrossRef]
45. Mulet, S.; Rio, M.H.; Etienne, H.; Artana, C.; Cancet, M.; Dibarboure, G.; Feng, H.; Husson, R.; Picot, N.; Provost, C.; et al. The new CNES-CLS18 global mean dynamic topography. *Ocean Sci.* **2021**, *17*, 789–808. [CrossRef]
46. U.S. Integrated Ocean Observing System. *Manual for Real-Time Quality Control of In-Situ Current Observations*; Version 2.1; U.S. Department of Commerce, National Oceanic and Atmospheric Administration, National Ocean Service, Integrated Ocean Observing System: Silver Spring, MD, USA, 2019. [CrossRef]
47. Min, Y.; Jeong, J.-Y.; Jang, C.J.; Lee, J.; Jeong, J.; Min, I.-K.; Shim, J.-S.; Kim, Y.S. Quality control of observed temperature time series from the Korea Ocean Research Stations: Preliminary application of ocean observation initiative's approach and its limitation. *Ocean Polar Res.* **2020**, *42*, 195–210. [CrossRef]
48. Leaman, K.D.; Sanford, T.B. Vertical energy propagation of inertial waves: A vector spectral analysis of velocity profiles. *J. Geophys. Res.* **1975**, *80*, 1975–1978. [CrossRef]
49. Lee, K.; Nam, S.; Kim, Y.-G. Statistical characteristics of east sea mesoscale eddies detected, tracked, and grouped using satellite altimeter data from 1993 to 2017. *Sea* **2019**, *24*, 267–281. [CrossRef]
50. Torrence, C.; Compo, G.P. A practical guide to wavelet analysis. *Bull. Am. Meteorol. Soc.* **1998**, *79*, 61–78. [CrossRef]
51. de Boyer Montégut, C. Mixed layer depth over the global ocean: An examination of profile data and a profile-based climatology. *J. Geophys. Res.* **2004**, *109*, C12003. [CrossRef]
52. Pollard, R.T.; Millard Jr, R. Comparison between observed and simulated wind-generated inertial oscillations. *Deep-Sea Res. Oceanogr. Abstr.* **1970**, *17*, 813–816. [CrossRef]
53. D'Asaro, E.A. The energy flux from the wind to near-inertial motions in the surface mixed layer. *J. Phys. Oceanogr.* **1985**, *15*, 1043–1059. [CrossRef]
54. Park, J.J.; Kim, K.; Schmitt, R.W. Global distribution of the decay timescale of mixed layer inertial motions observed by satellite-tracked drifters. *J. Geophys. Res. Oceans* **2009**, *114*, C11010. [CrossRef]
55. Kawaguchi, Y.; Wagawa, T.; Igeta, Y. Near-inertial internal waves and multiple-inertial oscillations trapped by negative vorticity anomaly in the central Sea of Japan. *Prog. Oceanogr.* **2020**, *181*, 102240. [CrossRef]
56. Chiswell, S.M. Deep equatorward propagation of inertial oscillations. *Geophys. Res. Lett.* **2003**, *30*, 1533. [CrossRef]
57. Okubo, A. Horizontal dispersion of floatable particles in the vicinity of velocity singularities such as convergences. *Deep-Sea Res. Oceanogr. Abstr.* **1970**, *17*, 445–454. [CrossRef]
58. Chavanne, C.P.; Firing, E.; Ascani, F. Inertial oscillations in geostrophic flow: Is the inertial frequency shifted by $\zeta/2$ or by ζ? *J. Phys. Oceanogr.* **2012**, *42*, 884–888. [CrossRef]
59. Shcherbina, A.Y.; Talley, L.D.; Firing, E.; Hacker, P. Near-surface frontal zone trapping and deep upward propagation of internal wave energy in the Japan/East Sea. *J. Phys. Oceanogr.* **2003**, *33*, 900–912. [CrossRef]
60. Song, H.; Jeon, C.; Chae, J.Y.; Lee, E.J.; Lee, K.N.; Takayama, K.; Choi, Y.; Park, J.H. Effects of typhoon and mesoscale eddy on generation and distribution of near-inertial wave energy in the East Sea. *Sea* **2020**, *25*, 55–66. [CrossRef]
61. Nam, S.; Kim, D.; Kim, H.R.; Kim, Y.G. Typhoon-induced, highly nonlinear internal solitary waves off the east coast of Korea. *Geophys. Res. Lett.* **2007**, *34*, L01607. [CrossRef]
62. Furuichi, N.; Hibiya, T.; Niwa, Y. Model-predicted distribution of wind-induced internal wave energy in the world's oceans. *J. Geophys. Res. Oceans* **2008**, *113*, C09034. [CrossRef]

63. Zhai, X.; Greatbatch, R.J.; Eden, C.; Hibiya, T. On the loss of wind-induced near-inertial energy to turbulent mixing in the upper ocean. *J. Phys. Oceanogr.* **2009**, *39*, 3040–3045. [CrossRef]
64. Jin, H.; Park, Y.G.; Pak, G.; Kim, Y.H. On the persistence of warm eddies in the East Sea. *Sea* **2019**, *24*, 318–331. [CrossRef]

MDPI
St. Alban-Anlage 66
4052 Basel
Switzerland
Tel. +41 61 683 77 34
Fax +41 61 302 89 18
www.mdpi.com

Journal of Marine Science and Engineering Editorial Office
E-mail: jmse@mdpi.com
www.mdpi.com/journal/jmse

www.ingramcontent.com/pod-product-compliance
Ingram Content Group UK Ltd.
Pitfield, Milton Keynes, MK11 3LW, UK
UKHW060124010826
14089UKWH00006B/147

9 783036 542133